Springer Theses

Recognizing Outstanding Ph.D. Research

Aims and Scope

The series "Springer Theses" brings together a selection of the very best Ph.D. theses from around the world and across the physical sciences. Nominated and endorsed by two recognized specialists, each published volume has been selected for its scientific excellence and the high impact of its contents for the pertinent field of research. For greater accessibility to non-specialists, the published versions include an extended introduction, as well as a foreword by the student's supervisor explaining the special relevance of the work for the field. As a whole, the series will provide a valuable resource both for newcomers to the research fields described, and for other scientists seeking detailed background information on special questions. Finally, it provides an accredited documentation of the valuable contributions made by today's younger generation of scientists.

Theses are accepted into the series by invited nomination only and must fulfill all of the following criteria

- They must be written in good English.
- The topic should fall within the confines of Chemistry, Physics, Earth Sciences, Engineering and related interdisciplinary fields such as Materials, Nanoscience, Chemical Engineering, Complex Systems and Biophysics.
- The work reported in the thesis must represent a significant scientific advance.
- If the thesis includes previously published material, permission to reproduce this must be gained from the respective copyright holder.
- They must have been examined and passed during the 12 months prior to nomination.
- Each thesis should include a foreword by the supervisor outlining the significance of its content.
- The theses should have a clearly defined structure including an introduction accessible to scientists not expert in that particular field.

Indexed by zbMATH.

More information about this series at http://www.springer.com/series/8790

Louise Desplat

Thermal Stability of Metastable Magnetic Skyrmions

Doctoral Thesis accepted by
University of Glasgow, Glasgow,
United Kingdom

Springer

Author
Dr. Louise Desplat
Institut de Physique et Chimie
des Matériaux de Strasbourg
CNRS and Université de Strasbourg
Strasbourg, France

Supervisors
Dr. Joo-Von Kim
Centre for Nanoscience
and Nanotechnology
CNRS Université Paris-Saclay
Palaiseau, France

Prof. Robert L. Stamps
Department of Physics and Astronomy
University of Manitoba
Manitoba, Canada

ISSN 2190-5053 ISSN 2190-5061 (electronic)
Springer Theses
ISBN 978-3-030-66025-3 ISBN 978-3-030-66026-0 (eBook)
https://doi.org/10.1007/978-3-030-66026-0

This Springer imprint is published by the registered company Springer Nature Switzerland AG
The registered company address is: Gewerbestrasse 11, 6330 Cham, Switzerland

So comes snow after fire, and even dragons have their endings!

J. R. R. Tolkien,
The Hobbit or There And Back Again.

Supervisors' Foreword

One of the most exciting developments in magnetism has been the physical realization of nontrivial topological structures called "skyrmions". At the time when this work commenced, the properties of these textured spin excitations had not been examined, and experiments were just beginning to demonstrate complex and often non-intuitive behaviours for skyrmions and curious, related objects. Moreover, methods were being rapidly developed for designing and fabricating new material systems in which specific properties of skyrmions could be engineered. Essential to understanding several of the puzzles presented by experiments, and underlying all of the potential applications was the need to understand and model the out-of-equilibrium behaviour of these mobile spin textures.

At the forefront of this rapidly developing context, the motivation of this thesis was to construct a theoretical framework for understanding lifetimes and decay rates for spin textures in low-dimensional structures with chiral interactions capable of supporting complex magnetic configurations such as skyrmions and anti-skyrmions. It involved tackling a thorny problem in magnetism, namely the quantitative evaluation of the Arrhenius rate equation for a system with a multi-dimensional phase space. This formulation expresses the rate as the product of a prefactor, commonly called the "attempt frequency", and an exponential function of the activation energy. The simplicity of this formulation, however, belies the complexity of the physical processes at play, as Dr. Louise Desplat demonstrates with great clarity and depth in her work.

The thesis begins with the description of the first of two complex and subtle theoretical and computational approaches employed for probing the question of skyrmion lifetimes. It involves a generalization of Kramer's framework to multi-dimensional phase spaces, due to Langer, which is most appropriate in the intermediate-to-high damping limit. This approach provides a framework for relating the spectra of eigen-excitations at the stable and saddle points to the Arrhenius prefactor, which was then applied to the problem of skyrmion collapse in different scenarios. On this, Louise makes the fascinating observation that strong entropic contributions can be the dominant factor in first passage rates, making knowledge of energy barriers alone insufficient to understand and predict lifetimes.

This observation is confirmed through the use of a second numerical technique, "forward flux sampling", which was first developed for chemical rate processes and recently extended to systems of interacting magnetic spins. While Langer's approach can prove unwieldy and relies on strong assumptions, forward flux sampling is free of assumptions and is based instead on path sampling of Langevin spin dynamics to estimate the probability of rare events. Here, Louise provides deep theoretical insight by providing a crucial comparison that helps to establish the validity of the forward flux computational method for further application to understanding and predicting complex dynamics in magnetic spin systems.

The final part of this thesis reveals extremely interesting consequences for thermal stability of topological excitations in systems with frustration. Here, a key fundamental understanding was obtained through studies of annihilation mechanisms for low-order skyrmions in frustrated systems. In this final work, mechanisms mediated by meron and antimeron pair production were identified and examined in detail. The work reflects a level of originality and insight that is exceptional for a doctoral thesis. We certainly enjoyed reading the dissertation and gained a better appreciation of the subtler aspects of the stochastic processes at play. We hope the reader will too.

Winnipeg, Canada
Palaiseau, France
December 2020

Prof. Robert L. Stamps
Dr. Joo-Von Kim

Abstract

Magnetic skyrmions are two-dimensional, localized, particle-like, topologically non-trivial magnetic spin textures. In recent years, they have attracted a lot of interest as potential candidates for novel spintronics applications. Isolated skyrmions are metastable excitations of the ferromagnetic ground state. They are separated from it by an activation energy, which may be overcome at a finite temperature under the effect of thermal fluctuations. In this thesis, we study the thermal stability of metastable magnetic skyrmions on the two-dimensional square lattice, for which we use an atomistic spin model.

This task is firstly carried out via a numerical implementation of Langer's statistical theory for the decay of metastable states. The paths of minimum energy that lead to the skyrmion annihilation are computed via the geodesic nudged elastic band method. The transition state at the barrier top, which is a saddle point (SP), is precisely identified by a climbing image algorithm. We focus on chiral magnetic skyrmions and we look at two types of annihilation mechanisms: collapse, in which the skyrrmion progressively shrinks in size until it annihilates, and escape through a boundary. We find that the thermally significant modes are the modes localized to the skyrmion, in contrast to the rest of the collective spin-wave modes, which extend to the entire lattice and contribute weakly. Important variations of the attempt frequency over several orders of magnitude are found, depending on the mechanism and on the value of the external magnetic field. They originate from strong entropic effects which come from the difference in configurational entropy between the metastable skyrmion state and the saddle point. In the cases we studied, the configurational entropy decreases at the SP ($\Delta S < 0$), which results in lowered attempt frequencies, and enhanced skyrmion stability. We refer to this effect as entropic narrowing in the SP region. The strong entropic contribution mainly originates from the skyrmion's internal modes and is generally more pronounced for collapse mechanisms.

Next, we use forward flux sampling to compute skyrmion collapse rates as a function of the applied field and compare them with the previous results from Langer's theory. This is an important step because the use of Langer's theory is based on many assumptions. We obtain a good agreement between both methods,

Publications related to the thesis

1. L. Desplat, C. Vogler, J.-V. Kim, R. L. Stamps, and D. Suess. Path sampling for lifetimes of metastable magnetic skyrmions and direct comparison with Kramers' method. *Physical Review B*, 101(6):060403(R), 2020.
2. L. Desplat, J.-V. Kim, and R. L. Stamps. Paths to annihilation of first- and second-order (anti)skyrmions via (anti)meron nucleation on the frustrated square lattice. *Physical Review B*, 99(17):174409, 2019.
3. L. Desplat, D. Suess, J.-V. Kim, and R. L. Stamps. Thermal stability of metastable magnetic skyrmions: Entropic narrowing and significance of internal eigenmodes. *Physical Review B*, 98:134407, 2018.

Acknowledgements

I shall begin by thanking Prof. Robert Stamps who supervised me in Glasgow for the first 2 years of my Ph.D. Bob was always kind and patient, and I sincerely enjoyed our scientific discussions. I also acknowledge Prof. Stephen McVitie who took care of the administrative side of my studies in Glasgow after Bob's departure. I thank the Materials and Condensed Matter Physics group in Glasgow for welcoming me for 2 years.

A most valuable collaboration over the course of my thesis has been with Prof. Dieter Suess from the University of Vienna, who welcomed me in his group for two consecutive years and almost 4 months in total. It was really instructional to discuss the stability of skyrmions with him, and I benefited a lot from his input. I later also worked with Dr. Christoph Vogler who offered some guidance for my implementation of forward flux sampling, for which I am grateful.

I am deeply indebted to Dr. Joo-Von Kim who kindly welcomed me in his group at the Centre for Nanoscience and Nanotechnology (C2N) at the University of Paris-Saclay for the last 2 years of my Ph.D., and took on the role of my main supervisor. Joo-Von's meticulous comments, insightful suggestions and warm encouragements were an invaluable help to me. Many thanks as well to everyone in NOMADE for welcoming me into their team.

As my Ph.D. grant was part of the MAGicSky consortium, I would like to thank the scientific coordinator, Prof. Vincent Cros, and all the members of the consortium who provided great learning opportunities during the project meetings. In particular, I thank Dr. Bertrand Dupé who welcomed me at the Christian Albrechts University of Kiel for a week, and who also offered me some valuable suggestions during his visits to the C2N. I extend my sincere thanks to Prof. Stefan Blügel for inviting me to give a seminar in FZ Jülich, and for the great discussions, I was able to have with him and the other members of the Peter Grünberg Institute. In particular, I thank Dr. Gideon Müller some insightful discussions. I also thank Dr. Pavel Bessarab from the University of Iceland for his useful remarks and for kindly answering questions via email. Thank you as well to Dr. Gerard Duff, formerly from University College Dublin for kindly clarifying for some points about his Ph.D. work.

I thank my thesis Examiners, Dr. Donald MacLaren from the University of Glasgow, and Prof. Gino Hrkac from the University of Exeter, as well as the Convener, Dr. Laura Clark.

Lastly, I thank my family for their support and my mother for her unwavering support and encouragement through the years.

This Ph.D. was funded by the Horizon 2020 Framework Programme of the European Commission, under Grant Agreement No. 665095 (MAGicSky), and the Agence Nationale de la Recherche under Contract No. ANR-17-CE24-0025 (TOPSKY).

Contents

Abbreviations

CI	Climbing image
DFT	Density functional theory
DMI	Dzyaloshinskii-Moriya interaction
FFS	Forward flux sampling
FM	Ferromagnetic
FPE	Fokker-Planck equation
GNEB	Geodesic nudged elastic band
IHD	Intermediate-to-high damping
im.	Image
LLG	Landau-Lifshitz-Gilbert
MEP	Minimum energy path
PMA	Perpendicular magnetic anisotropy
SOC	Spin-orbit coupling
SP	Saddle point
SP-STM	Spin-polarized scanning tunneling microscopy
SS	Spin spiral
TEM	Transmission electron microscopy
TST	Transition state theory

Chapter 1
Introduction

Louise Desplat

Historically, a soliton phenomenon was first described in 1834 by civil engineer John Scott Russell, in the Edinburgh and Glasgow Union Canal in Scotland [1]. He observed a propagating water wave with great stability that travelled through a large distance without loss of shape or velocity, and named it a "wave of translation". He later reproduced the phenomenon in a water tank and, apart from their stability, described the following properties of solitons: (i) their speed depends on their size and their width on the water depth; (ii) waves do not merge, and instead a bigger and faster wave travels through a smaller and slower one; (iii) a wave too big for the water depth separates into two smaller waves. More generally, solitons correspond to localized solutions to a nonlinear wave equation, and are the result of a balance between dispersive effects and nonlinearities [2]. One example is the sine-Gordon equation, which we give in (1+1) dimensions for the scalar field $\phi(x, t)$ [3],

$$\partial_t^2 \phi - \partial_x^2 \phi + \sin\phi = 0. \tag{1.1}$$

Equation (1.1) admits solitons as some of its solutions. Solitons such as waves in water are intrinsically dynamical objects, and simply relax back to the uniform state in the static case. By contrast, in systems with strong couplings between the microscopic entities, as is the case in magnetic materials, a class of solitons known as topological solitons can be statically stable due to topological constraints. Examples of topological solitons in magnetism, also known as topological defects, include domain walls, vortices, skyrmions, and bubbles [4]. We show some typical spin maps of the aforementioned structures in Fig. 1.1.

L. Desplat (✉)
Institut de Physique et Chimie des Matériaux de Strasbourg, CNRS and Université de Strasbourg, Strasbourg, France
e-mail: louise.desplat@ipcms.unistra.fr

L. Desplat, *Thermal Stability of Metastable Magnetic Skyrmions*, Springer Theses,
https://doi.org/10.1007/978-3-030-66026-0_1

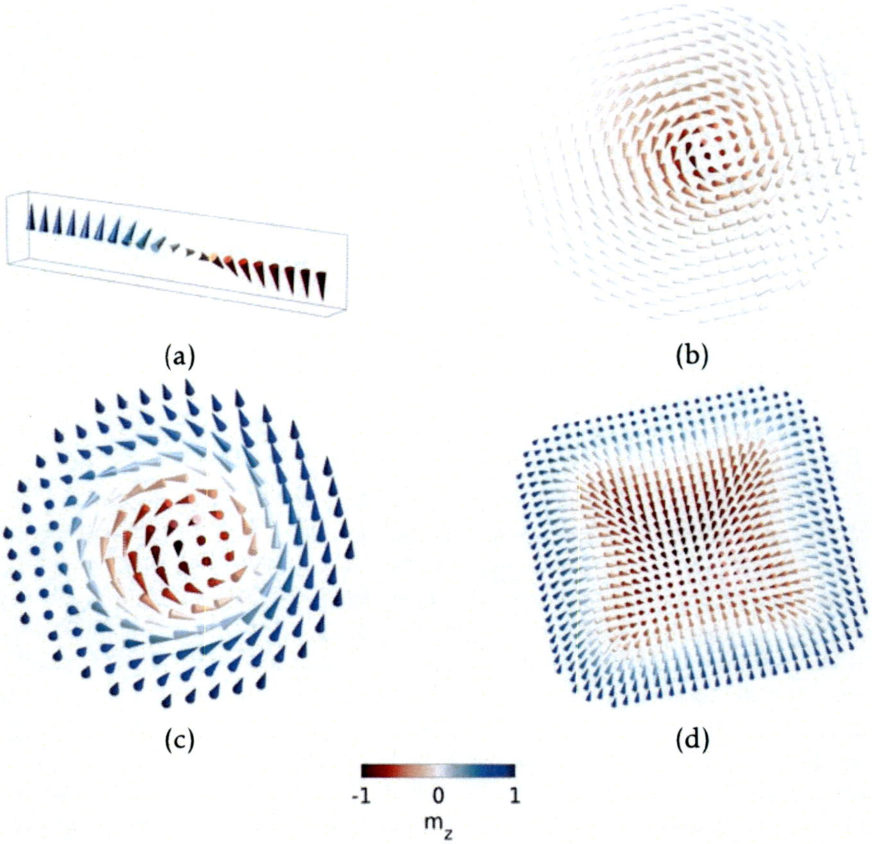

Fig. 1.1 Spin maps of some topological solitons in magnetism. **a** A (Bloch) domain wall in a chain of spins. The magnetization undergoes a π-rotation, from one out-of-plane orientation to the other. **b** A magnetic vortex. The magnetization goes from out of plane in the centre, to in plane far from the centre, and thus undergoes a total π-rotation along a radial cross-section. **b** A (Bloch) magnetic skyrmion. The magnetization rotates from one out-of-plane orientation far from the centre, to the opposite out-of-plane orientation in the centre, resulting in a total 2π-rotation along a radial cross-section. **c** A bubble domain. Similar to a skyrmion, the magnetization rotates by 2π along a radial cross section, but the profile is different, with a more spatially extended core. It follows that domain walls and vortices carry a half-integer topological charge, while skyrmions and bubbles carry an integer charge

The term "skyrmion" originated from baryonic field theories, and initially designated a type of solitonic solutions derived by Skyrme [5] within his model for low energy dynamics between mesons and baryons. The term "magnetic skyrmion" was later introduced into the field of condensed matter physics to designate so-called "chiral magnetic vortices" theoretically studied by Bogdanov et al. in the 1990s [6–8]. Magnetic skyrmions are two-dimensional, non-collinear magnetic spin textures which are localized in space (Fig. 1.2a). Most commonly, the existence of skyrmion solutions in magnets is made possible by the antisymmetric Dzyaloshinskii-Moriya interaction (DMI) [9, 10] which favours non-collinear spin textures in systems lacking inversion symmetry. DMI exists in non-centrosymmetric magnets, and can also be induced in centrosymmetric crystals if the symmetry is broken by an external mechanism. In particular, interfacial DMI emerges in thin magnetic films due to the breaking of the inversion symmetry at the interface [11, 12]. Magnetic skyrmions stabilized by the DMI are known as chiral magnetic skyrmions, because the DMI selects a preferred chirality—or handedness—of the spin texture. However, they may be stabilized in centrosymmetric magnets where long-range dipole-dipole couplings compete with the dominant short-range isotropic exchange term (skyrmion bubbles) [13–18], or by introducing frustration in the exchange [19–24].

A skyrmion lattice was first detected in the bulk of the B20 chiral magnet MnSi in reciprocal space, via neutron scattering [25]. Shortly after, Yu et al. reported the first real space observation of a skyrmion lattice in a thin film of $Fe_{0.5}Co_{0.5}$ by means of Lorentz transmission electron microscopy [26]. These observations initiated a strong regain of interest in the community, and the existence of skyrmions was then confirmed in many other types of materials [27–29], including thin films [30, 31]. Similarly to domain walls [14], which have been used as information carriers in racetrack memories [32], magnetic skyrmions can be moved with spin torques associated with spin polarized currents [33, 34], and have therefore been envisioned for potential spintronics applications in data storage and logic devices [35–39]. Contrary to individual π-domain walls, skyrmions are localized in two dimensions, and therefore leave the magnetization field intact far away from the centre. Their equilibrium sizes are determined by the strength of the different competing energy terms, and it has been proposed that skyrmion with sizes down to the nanometer scale may be stabilized by adequately tuning material parameters [40, 41]. Additionally, they exhibit a particle-like behaviour which finds its root in the topology, and is independent of the underlying stabilizing mechanism [21, 42]. For technological applications, it it necessary to design structures with great stability and robustness against external effects and deformations. Because they are classified as "topologically stable", and for all the reasons mentioned above, skyrmions are compelling candidates to be used as movable information bits in future spintronics devices.

Potential stability arguments for skyrmions are naturally found in the concepts of topology (or homotopy), and winding number. Further details on the content of the following paragraph can be found in Ref. [4]. Topologically nontrivial textures may be seen as a knot in a rope, which cannot be untied without using the extremities, or by cutting the rope. The stability of such textures thus stems from their resistance against deformation. Configurations are said to be topologically (or homotopically) equivalent if it is possible to continuously transform one into the other without introducing point defects. Topological stability arises if a structure is topologically nontrivial, and therefore cannot be continuously deformed into the topologically trivial uniform state. Mutually deformable configurations form a class of mappings,

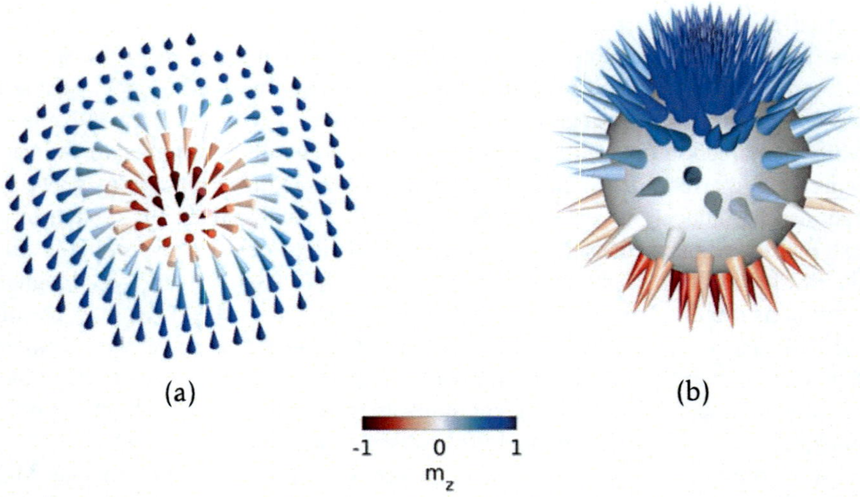

Fig. 1.2 **a** Néel "hedgehog" skyrmion and **b** stereographic projection onto the 2-sphere S^2

$f : S^n \rightarrow X$ of the $n-$sphere S^n (the "base manifold") into the order-parameter space X (the "target manifold"), and yield equivalence—or homotopy—classes $\pi_n(X)$ [4]. The winding number, or degree of mapping, deg f, counts the number of times the target manifold X wraps around the base manifold S^n. Formally, it provides a bijection of the homotopy classes into the integers $\mathbb{Z}$, and may be used to classify homotopy classes. That is to say, configurations with different winding numbers do not pertain to the same homotopy class and are therefore not topologically equivalent, and so they cannot be continuously transformed into each other.

In magnetism, we are concerned with the mappings provided by the magnetization field $\mathbf{m}(\mathbf{r})$. In an infinite system, two magnetic configurations are said to be topologically (or homotopically) equivalent if it is possible to continuously transform one into the other without having to overcome an infinite energy barrier [42]. The uniformly magnetized state, also referred to as the ferromagnetic (FM) state, has a vanishing winding number, which makes it topologically trivial. On the other hand, topological defects such as magnetic skyrmions are topologically non-trivial. In isotropic spin systems with two spatial dimensions, the order parameter space is the 2-sphere S^2. As we assume that the magnetization takes a constant value at infinity, the base manifold is the physical space $\mathbb{R}^2$ with infinity compactified to a point, which may be stereographically projected onto S^2 (Fig. 1.2b). We then have the map $f : S^2 \rightarrow S^2$, which corresponds to the homotopy class $\pi_2(S^2) = \mathbb{Z}$ [4]. Skyrmions possess an integer winding number, and consequently cannot be continuously transformed into the uniform state. For this reason, they are classified as "topologically stable". Nevertheless, this classification corresponds to an idealized viewpoint. The notion of topological stability implies that the magnetization varies slowly with respect to the lattice, and thus may be seen as a continuous vector field $\mathbf{m}(\mathbf{r}, t)$. In physical systems, the energy barriers are not infinite [42] because (i) the magnetization is defined on a discrete lattice and thus may, in some cases, vary on the scale of the lattice parameter, and (ii) samples are finite in size and so topological defects may be injected or ejected at the edges. The finite energy barriers imply that isolated skyrmions which exist as excitations of the ferromagnetic ground state are, in fact, metastable states with a finite lifetime, τ. At non-zero temperature, there is a finite probability that they eventually annihilate under the effect of thermal fluctuations, and the uniformly magnetized configuration is recovered. In order for skyrmions to be used in applications, one must be able to, firstly, predict their average lifetime, and, secondly, understand it in order to potentially tune it to achieve high stability in room-temperature conditions. For data storage applications, the technical requirement is an average lifetime of ten years at room temperature [43]. We are thus faced with the problem of estimating the rate at which a metastable state decays.

The rate of decay of metastable states is an ubiquitous problem in physics. Thermal activation processes across an energy barrier are found within fields as diverse as solid state physics (Josephson tunnelling junctions), chemical reactions, electrical circuit theory (phase-locked loops), laser physics, and magnetization switching in ferromagnets [44, 45]. It can be described as originating from the nucleation and growth of some characteristic disturbance within the system [46]. Amongst the most well known examples of this phenomenon, one can cite the case of a supersaturated vapour, which can exist as a metastable state for a long time, until the nucleation of a large enough water droplet leads to the condensation of the entire system. Similarly, in a ferromagnet, a magnetization reversal is initiated by a cluster of reversed spins. The rate of such thermally activated processes may be generally described by the Arrhenius law [47],

$$k = f_0 e^{-\Delta E / k_B T}, \tag{1.2}$$

in which ΔE is the internal energy barrier, and the pre-exponential factor f_0, commonly referred to as the attempt frequency, corresponds to a fundamental fluctuation rate. Equation (1.2) laid the foundations for modern reaction rate theory. The mean lifetime of a metastable state may be taken as the inverse of its decay rate, $\tau = k^{-1}$. To calculate the rate, the following approaches are possible. On the one hand, a treatment of the different terms on the right-hand side of Eq. (1.2) may be achieved by applying a form of reaction rate theory [44, 47]. This requires the evaluation of the activation energy ΔE, and of the rate prefactor f_0. Alternatively, a total rate k may be computed through numerical integration of the system dynamics. In the rest of this discussion, we focus on magnetic systems.

Historically, the rate of magnetization reversal in single-domain magnetic particles has been extensively studied [48]. Since their magnetization can be described by a macrospin with only two degrees of freedom, they are well suited for an analytical treatment. In the case of inhomogeneous transitions, thermally activated magnetization reversals were treated analytically by Braun in nanowires [49], by Loxley and Stamps in soft/hard nanowires [50], and by Visscher et al. in nanowires with graded anisotropy [51]. For more complex systems, one typically has to rely on numerical methods. A systematic application of reaction rate theory may be realized in two main steps. First, a path finding scheme, such as the nudged elastic band method

(NEB) [52–56] or the string method [57, 58], may be used to relax minimum energy paths separating two energy minima, and identify the saddle point (SP) along the path. A minimum energy path constitutes a path of highest statistical weight, and therefore corresponds to a particular mechanism of a given transition. After that, the rate prefactor needs to be evaluated. This calculation tends to be tedious, and has rarely been carried out [59–61]. Instead, estimating the stability of magnetic structures is often synonymous with evaluating internal energy barriers, while assuming a typical value of the prefactor in the gigahertz range [62–65].

An alternative approach lies in the numerical integration of the dynamical equations of motion at finite temperature, i.e., Langevin simulations. In magnetic spin systems, the magnetization dynamics is governed by the Landau-Lifshitz-Gilbert (LLG) equation. For structures with lifetimes well above the nanosecond range, transitions become *rare events* on the timescale of the simulation, due to the small timesteps ($dt \sim 10^{-15}$ s) required to resolve the precessional dynamics [66]. Therefore, the "brute-force" method that relies on direct Langevin simulations becomes unrealistic. One may resort to temperature-accelerated dynamics [67, 68], but such simulations reproduce the low-temperature behaviour only under the assumption that reaction rate theory holds. For structures whose lifetime extends to the scale of years, temperatures well above the Curie temperature are required in order to obtain the necessary acceleration, and therefore a problem of consistency arises if one aims to study ordered spin states. In this case, a better suited class of methods are path sampling methods [69–75]. In particular, forward flux sampling (FFS) [72–75] may be used to compute rate constants of rare events even in systems that are out of equilibrium. Forward flux sampling was shown to be much more computationally efficient than brute force simulations [73, 75]. It has, for instance, been successfully applied by Vogler et al. [76, 77] to obtain switching rates in magnetic microstructures.

The aim of this thesis is to study the thermal stability of metastable magnetic skyrmions in two-dimensional systems. This problem is difficult to solve analytically and requires numerical tools. In micromagnetics simulations, where the magnetization is treated as a continuous vector field, and within a three-dimensional physical space, skyrmions typically decay via the formation of a Bloch point, a topological singularity where the continuity of magnetism is broken [78]. While Bloch points do not strictly exist in two dimensions, equivalent processes to the three dimensional ones have been reported (see discussion in [78] and subsequent refs.). The use of atomistic simulations is therefore necessary in order to avoid a mesh-size dependency of the activation rates [79]. The annihilation rates are thus calculated on an atomistic spin lattice via two main methods. The first one is a form of reaction rate theory which corresponds to Kramers' method [80] for a multidimensional phase-space in the intermediate-to-high damping (IHD) regime, namely Langer's statistical theory of the decay of metastable states [46]. The second one is forward flux sampling.

The present thesis is organized as follows. In Chap. 2, we give the different terms in the magnetic Hamiltonian, and we discuss in detail two types of topological solitons: ferromagnetic walls and magnetic skyrmions. In Chap. 3, we present Langer's theory and how it is applied to magnetic spin systems. We then provide details of the different building blocks in the implementation of Langer's theory on the atomistic spin lattice. In particular, we present the geodesic nudged elastic band (GNEB) method, which allows us to relax minimum energy paths on the energy surface. After that, Langer's theory is applied to obtain annihilation rates of metastable magnetic skyrmions via collapse and escape processes in Chap. 4. In Chap. 5, we expose the principles behind the forward flux sampling method, and we apply it to compute collapse rates of metastable magnetic skyrmions in an infinite system. The obtained results are subsequently compared with those from Langer's theory. Next, in Chap. 6, we introduce frustration in the exchange and, in that context, we look at paths to annihilation of skyrmionic solutions in frustrated systems, and their stability is discussed. A general discussion of the results and some perspectives are, lastly, given in Chap. 7.

References

1. Russell JS (1844) Report on waves. In: 14th meeting of the British Association for the advancement of science, pp 90–311
2. Drazin PG, Johnson RS (1989) Solitons: an introduction. Cambridge University Press, Cambridge, United Kingdom
3. Rajaraman R (1982) Solitons and instantons: an introduction to solitons and instantons in quantum field theory. North-Holland, Amsterdam (Netherlands)
4. Zang J, Cros V, Hoffmann A (2018) Topology in magnetism. Springer series in solid-state sciences, vol 192. Springer Nature, Switzerland
5. Skyrme THR (1961) A non-linear field theory. Proc R Soc 260:127
6. Bogdanov A, Yablonskii D (1989) Thermodynamically stable "vortices" in magnetically ordered crystals. The mixed state of magnets. Zhurnal Éksperimental'noĭ i Teoreticheskoĭ Fiziki 95:182
7. Ivanov B, Stephanovich V, Zhmudskii A (1990) Magnetic vortices - the microscopic analogs of magnetic bubbles. J Magn Magn Mater 88(1–2):116–120
8. Bogdanov A, Hubert A (1994) Thermodynamically stable magnetic vortex states in magnetic crystals. J Magn Magn Mater 138(3):255–269
9. Dzyaloshinskii IE (1958) A thermodynamic theory of "weak" ferromagnetism of antiferromagnetics. J Phys Chem Solids 4(4):241–255
10. Moriya T (1960) Anisotropic superexchange interaction and weak ferromagnetism. Phys Rev 120(1):91–98
11. Fert A (1990) Magnetic and transport properties of metallic multilayers. In: Materials science forum, vol 59. Trans Tech Publications, pp 439–480
12. Crépieux A, Lacroix C (1998) Dzyaloshinsky-Moriya interactions induced by symmetry breaking at a surface. J Magn Magn Mater 182(3):341–349
13. Chikazumi S, Graham CD (2009) Physics of ferromagnetism. International series of monographs on physics, vol 94, 2nd edn. Oxford University Press
14. Malozemoff A, Slonczewski J (1979) Magnetic domain walls in bubble materials: advances in materials and device research, vol 1. Academic Press
15. Yu X, Mostovoy M, Tokunaga Y, Zhang W, Kimoto K, Matsui Y, Kaneko Y, Nagaosa N, Tokura Y (2012) Magnetic stripes and skyrmions with helicity reversals. Proc Natl Acad Sci 109(23):8856–8860
16. Finazzi M, Savoini M, Khorsand A, Tsukamoto A, Itoh A, Duo L, Kirilyuk A, Rasing T, Ezawa M (2013) Laser-induced magnetic nanostructures with tunable topological properties. Phys Rev Lett 110(17):177205

17. Nagao M, So Y-G, Yoshida H, Isobe M, Hara T, Ishizuka K, Kimoto K (2013) Direct observation and dynamics of spontaneous skyrmion-like magnetic domains in a ferromagnet. Nat Nanotechnol 8(5):325
18. Koshibae W, Nagaosa N (2016) Theory of antiskyrmions in magnets. Nat Commun 7:10542
19. Leonov A, Mostovoy M (2015) Multiply periodic states and isolated skyrmions in an anisotropic frustrated magnet. Nat Commun 6:8275
20. Okubo T, Chung S, Kawamura H (2012) Multiple-q states and the skyrmion lattice of the triangular-lattice Heisenberg antiferromagnet under magnetic fields. Phys Rev Lett 108(1):017206
21. Nagaosa N, Tokura Y (2013) Topological properties and dynamics of magnetic skyrmions. Nat Nanotechnol 8(12):899–911
22. Lin S-Z, Hayami S (2016) Ginzburg-Landau theory for skyrmions in inversion-symmetric magnets with competing interactions. Phys Rev B 93(6):064430
23. Rózsa L, Palotás K, Deák A, Simon E, Yanes R, Udvardi L, Szunyogh L, Nowak U (2017) Formation and stability of metastable skyrmionic spin structures with various topologies in an ultrathin film. Phys Rev B 95(9):094423
24. Zhang X, Xia J, Zhou Y, Liu X, Zhang H, Ezawa M (2017) Skyrmion dynamics in a frustrated ferromagnetic film and current-induced helicity locking-unlocking transition. Nat Commun 8(1):1717
25. Mühlbauer S, Binz B, Jonietz F, Pfleiderer C, Rosch A, Neubauer A, Georgii R, Böni P (2009) Skyrmion lattice in a chiral magnet. Science 323(5916):915–919
26. Yu X, Onose Y, Kanazawa N, Park J, Han J, Matsui Y, Nagaosa N, Tokura Y (2010) Real-space observation of a two-dimensional skyrmion crystal. Nature 465(7300):901
27. Tokunaga Y, Yu X, White J, Rønnow HM, Morikawa D, Taguchi Y, Tokura Y (2015) A new class of chiral materials hosting magnetic skyrmions beyond room temperature. Nat Commun 6:7638
28. Nayak AK, Kumar V, Ma T, Werner P, Pippel E, Sahoo R, Damay F, Rößler UK, Felser C, Parkin SS (2017) Magnetic antiskyrmions above room temperature in tetragonal Heusler materials. Nature 548(7669):561
29. Kanazawa N, Seki S, Tokura Y (2017) Noncentrosymmetric magnets hosting magnetic skyrmions. Adv Mater 29(25):1603227
30. Romming N, Hanneken C, Menzel M, Bickel JE, Wolter B, von Bergmann K, Kubetzka A, Wiesendanger R (2013) Writing and deleting single magnetic skyrmions. Science 341(6146):636–639
31. Heinze S, Von Bergmann K, Menzel M, Brede J, Kubetzka A, Wiesendanger R, Bihlmayer G, Blügel S (2011) Spontaneous atomic-scale magnetic skyrmion lattice in two dimensions. Nat Phys 7(9):713
32. Parkin SS, Hayashi M, Thomas L (2008) Magnetic domain-wall racetrack memory. Science 320(5873):190–194
33. Sampaio J, Cros V, Rohart S, Thiaville A, Fert A (2013) Nucleation, stability and current-induced motion of isolated magnetic skyrmions in nanostructures. Nat Nanotechnol 8(11):839
34. Woo S, Litzius K, Krüger B, Im M-Y, Caretta L, Richter K, Mann M, Krone A, Reeve RM, Weigand M et al (2016) Observation of room-temperature magnetic skyrmions and their current-driven dynamics in ultrathin metallic ferromagnets. Nat Mater 15(5):501
35. Fert A, Cros V, Sampaio J (2013) Skyrmions on the track. Nat Nanotechnol 8(3):152
36. Iwasaki J, Mochizuki M, Nagaosa N (2013) Current-induced skyrmion dynamics in constricted geometries. Nat Nanotechnol 8(10):742
37. Zhou Y, Ezawa M (2014) A reversible conversion between a skyrmion and a domain-wall pair in a junction geometry. Nat Commun 5:4652
38. Müller J (2017) Magnetic skyrmions on a two-lane racetrack. New J Phys 19(2):025002
39. Hsu P-J, Kubetzka A, Finco A, Romming N, von Bergmann K, Wiesendanger R (2017) Electric-field-driven switching of individual magnetic skyrmions. Nat Nanotechnol 12(2):123
40. Dupé B, Hoffmann M, Paillard C, Heinze S (2014) Tailoring magnetic skyrmions in ultra-thin transition metal films. Nat Commun 5:4030

41. Dupé B, Bihlmayer G, Böttcher M, Blügel S, Heinze S (2016) Engineering skyrmions in transition-metal multilayers for spintronics. Nat Commun 7:11779
42. Braun H-B (2012) Topological effects in nanomagnetism: from superparamagnetism to chiral quantum solitons. Adv Phys 61(1):1–116
43. Wild J, Meier TN, Pöllath S, Kronseder M, Bauer A, Chacon A, Halder M, Schowalter M, Rosenauer A, Zweck J et al (2017) Entropy-limited topological protection of skyrmions. Sci Adv 3(9):e1701704
44. Coffey W, Garanin D, McCarthy D (2001) Crossover formulas in the Kramers theory of thermally activated escape rates—application to spin systems. Adv Chem Phys 117:483–765
45. Risken H (1996) The Fokker-Planck equation. Springer series in synergetic, vol 18. Springer, Berlin, Heidelberg
46. Langer JS (1969) Statistical theory of the decay of metastable states. Ann Phys 54:258–275
47. Hänggi P, Talkner P, Borkovec M (1990) Reaction-rate theory: fifty years after Kramers. Rev Modern Phys 62(2):251
48. Coffey WT, Kalmykov YP (2012) Thermal fluctuations of magnetic nanoparticles: fifty years after Brown. J Appl Phys 112(12):121301
49. Braun H-B (1994) Kramers's rate theory, broken symmetries, and magnetization reversal. J Appl Phys 76(10):6310–6315
50. Loxley PN, Stamps RL (2006) Theory for nucleation at an interface and magnetization reversal of a two-layer nanowire. Phys Rev B 73(2):024420
51. Visscher PB, Zhu R (2012) Low-dimensionality energy landscapes: magnetic switching mechanisms and rates. Phys B: Condens Matter 407(9):1340–1344
52. Mills G, Jónsson H, Schenter GK (1995) Reversible work transition state theory: application to dissociative adsorption of hydrogen. Surf Sci 324(2–3):305–337
53. Jónsson H, Mills G, Jacobsen KW (1998) Nudged elastic band method for finding minimum energy paths of transitions. In: Classical and quantum dynamics in condensed phase simulations, pp 385–404
54. Dittrich R, Schrefl T, Suess D, Scholz W, Forster H, Fidler J (2002) A path method for finding energy barriers and minimum energy paths in complex micromagnetic systems. J Magn Magn Mater 250:12–19
55. Henkelman G, Uberuaga BP, Jónsson H (2000) A climbing image nudged elastic band method for finding saddle points and minimum energy paths. J Chem Phys 113(22):9901–9904
56. Bessarab PF, Uzdin VM, Jonsson H (2015) Method for finding mechanism and activation energy of magnetic transitions, applied to skyrmions and antivortex annihilation. Comput Phys Commun 196:335–347
57. Weinan E, Ren W, Vanden-Eijnden E (2002) String method for the study of rare events. Phys Rev B 66(5):052301
58. Weinan E, Ren W, Vanden-Eijnden E (2007) Simplified and improved string method for computing the minimum energy paths in barrier-crossing events. J Chem Phys 126(16):164103
59. Fiedler G, Fidler J, Lee J, Schrefl T, Stamps RL, Braun H, Suess D (2012) Direct calculation of the attempt frequency of magnetic structures using the finite element method. J Appl Phys 111(9):093917
60. Bessarab PF, Müller GP, Lobanov IS, Rybakov FN, Kiselev NS, Jónsson H, Uzdin VM, Blügel S, Bergqvist L, Delin A (2018) Lifetime of racetrack skyrmions. Sci Rep 8(1):1–10
61. Desplat L, Suess D, Kim J-V, Stamps RL (2018) Thermal stability of metastable magnetic skyrmions: entropic narrowing and significance of internal eigenmodes. Phys Rev B 98:134407
62. Weller D, Moser A (1999) Thermal effect limits in ultrahigh-density magnetic recording. IEEE Trans Magn 35(6):4423–4439
63. Chen E, Apalkov D, Diao Z, Driskill-Smith A, Druist D, Lottis D, Nikitin V, Tang X, Watts S, Wang S et al (2010) Advances and future prospects of spin-transfer torque random access memory. IEEE Trans Magn 46(6):1873–1878
64. Lederman M, Schultz S, Ozaki M (1994) Measurement of the dynamics of the magnetization reversal in individual single-domain ferromagnetic particles. Phys Rev Lett 73(14):1986

65. Cortés-Ortuño D, Wang W, Beg M, Pepper RA, Bisotti M-A, Carey R, Vousden M, Kluyver T, Hovorka O, Fangohr H (2017) Thermal stability and topological protection of skyrmions in nanotracks. Sci Rep 7(1):4060
66. García-Palacios JL, Lázaro FJ (1998) Langevin-dynamics study of the dynamical properties of small magnetic particles. Phys Rev B 58:14937–14958 Dec
67. Voter AF (1997) Hyperdynamics: accelerated molecular dynamics of infrequent events. Phys Rev Lett 78(20):3908
68. Xue J, Victora R (2000) Micromagnetic predictions for thermally assisted reversal over long time scales. Appl Phys Lett 77(21):3432–3434
69. Dellago C, Bolhuis PG, Csajka FS, Chandler D (1998) Transition path sampling and the calculation of rate constants. J Chem Phys 108(5):1964–1977
70. Bolhuis PG, Chandler D, Dellago C, Geissler PL (2002) Transition path sampling: throwing ropes over rough mountain passes, in the dark. Annu Rev Phys Chem 53(1):291–318
71. van Erp TS, Moroni D, Bolhuis PG (2003) A novel path sampling method for the calculation of rate constants. J Chem Phys 118(17):7762–7774
72. Allen RJ, Warren PB, ten Wolde PR (2005) Sampling rare switching events in biochemical networks. Phys Rev Lett 94:018104
73. Allen RJ, Frenkel D, ten Wolde PR (2006) Simulating rare events in equilibrium or nonequilibrium stochastic systems. J Chem Phys 124(2):024102
74. Allen RJ, Frenkel D, ten Wolde PR (2006) Forward flux sampling-type schemes for simulating rare events: efficiency analysis. J Chem Phys 124(19):194111
75. Allen RJ, Valeriani C, ten Wolde PR (2009) Forward flux sampling for rare event simulations. J Phys Condens Matter 21(46):463102
76. Vogler C, Bruckner F, Bergmair B, Huber T, Suess D, Dellago C (2013) Simulating rare switching events of magnetic nanostructures with forward flux sampling. Phys Rev B 88(13):134409
77. Vogler C, Bruckner F, Suess D, Dellago C (2015) Calculating thermal stability and attempt frequency of advanced recording structures without free parameters. J Appl Phys 117(16):163907
78. Rohart S, Miltat J, Thiaville A (2016) Path to collapse for an isolated Néel skyrmion. Phys Rev B 93(21):214412
79. Suess D, Vogler C, Bruckner F, Heistracher P, Abert C (2018) A repulsive skyrmion chain as a guiding track for a racetrack memory. AIP Adv 8(11):115301
80. Kramers HA (1940) Brownian motion in a field of force and the diffusion model of chemical reactions. Physica 7(4):284–304

Chapter 2
Topological Solitons in Magnetic Systems

In this chapter, we begin by introducing the different terms in the classical Hamiltonian model for ferromagnets in Sect. 2.1. In Sect. 2.2, we give some elements of theory for ferromagnetic domain walls, which we use to introduce the key notions of winding number and topology. Finally, in Sect. 2.3, we define magnetic skyrmions and present the different types of systems in which they can be stabilized.

2.1 The Classical Hamiltonian Model

A ferromagnet is characterized by a spontaneous magnetization associated with long range magnetic ordering. This originates from exchange interactions between atoms, which favour parallel orientations of neighbouring atomic spins, and constitute the dominant energy contribution in magnetic materials. The different terms in the energy of a magnetic system describe the interaction of the magnetic moments with an external field, the crystal lattice, as well as interactions with the other magnetic moments. These consist of short-range exchange interactions originating from quantum mechanics, and long-range magnetostatic interactions. Equilibrium magnetization structures are found by minimizing the total energy of a system, and correspond to the case when the total effective field exerts no torque on the magnetic vectors.

We consider an ensemble of N atoms on a lattice with lattice constant a (Fig. 2.1a). Each atom localized at site i carries a net magnetic moment that we write in the form of a classical vector $\mathbf{m}_i = \mathbf{M}_i / M_s$, in which M_s is the saturation magnetization and $\mathbf{m}_i$ is a unit vector. The magnetic moment is linked to the atomic spin $\mathbf{S}_i$ by the relation $\mathbf{M}_i = -g\mu_B \mathbf{S}_i$ [1]. In the previous expression, g is the Landé factor and μ_B is the Bohr magneton. In the rest of this thesis, we use the words spins and magnetic moments interchangeably. The total energy of the assembly of spins in its atomistic form is

$$E = E_{\mathrm{ex}} + E_{\mathrm{DM}} + E_Z + E_{\mathrm{an}}, \tag{2.1}$$

L. Desplat, *Thermal Stability of Metastable Magnetic Skyrmions*, Springer Theses,
https://doi.org/10.1007/978-3-030-66026-0_2

where E_{ex} is the isotropic exchange energy, E_{DM} is the DMI exchange energy, E_Z is the Zeeman energy, and E_{an} is the anisotropy energy. Note that we neglect dipolar couplings, although they are discussed later in this chapter.

Within the framework of micromagnetics theory [2], we assume only small angular variations between the orientations of neighbouring spins, which allows us to write the magnetization as a continuous vector field, $\mathbf{m} = \mathscr{M}/\mathscr{M}_s$, with norm $\mathscr{M}_s$. In the rest of this work, micromagnetics quantities are symbolized with cursive letters—with the exception of the spin stiffness, A—while regular symbols represent atomistic quantities. The micromagnetics Hamiltonian reads

$$\mathscr{E} = \mathscr{E}_{ex} + \mathscr{E}_{DM} + \mathscr{E}_{Z} + \mathscr{E}_{an}, \tag{2.2}$$

where the indices have the same meaning as in Eq. (2.1).

In the rest of this section, we introduce the different terms in Eq. (2.1) within the classical Heisenberg model, as well as their corresponding expressions in Eq. (2.2) from micromagnetics theory. We give particular emphasis to the case of thin magnetic films with perpendicular magnetic anisotropy (PMA), as this is a category of systems which is currently the most envisioned for spintronics applications, as well as the type of system that was studied in the present thesis.

Let us define here the spherical coordinate system $(1, \theta, \phi)$ (Fig. 2.1b) which we shall use throughout this thesis. θ is the polar angle of the magnetization with the z-axis, and ϕ is the corresponding azimuth angle in the xy-plane, so that the unit magnetic vector $\mathbf{m} = (m_x, m_y, m_z)$ may be expressed as

$$\mathbf{m} = \left(\cos\phi\sin\theta, \sin\phi\sin\theta, \cos\theta\right). \tag{2.3}$$

2.1.1 Exchange Interactions

2.1.1.1 Isotropic Heisenberg Exchange

The isotropic exchange interaction originates from the overlap of electronic orbitals and was first proposed by Heisenberg in 1928 [3] to interpret the large molecular field observed in ferromagnets. A year later, Dirac [4] showed that the interaction energy between two neighbouring sites i,j can be expressed in its atomistic form as $-2J_{ij}\mathbf{m}_i \cdot \mathbf{m}_j$, where J_{ij} is the exchange integral between sites i and j. It decreases rapidly as a function of distance, which in turn explains the short-range nature of the interaction. Isotropic exchange is referred to as ferromagnetic if it favours parallel spins, i.e., if $J_{ij} > 0$, and antiferromagnetic if $J_{ij} < 0$, which favours anti-parallel spins. The interaction is symmetric upon interchanging i and j, and isotropic as it is independent of the direction of the change relative to the magnetization direction. The total isotropic exchange energy of a system of N interacting spins is expressed as

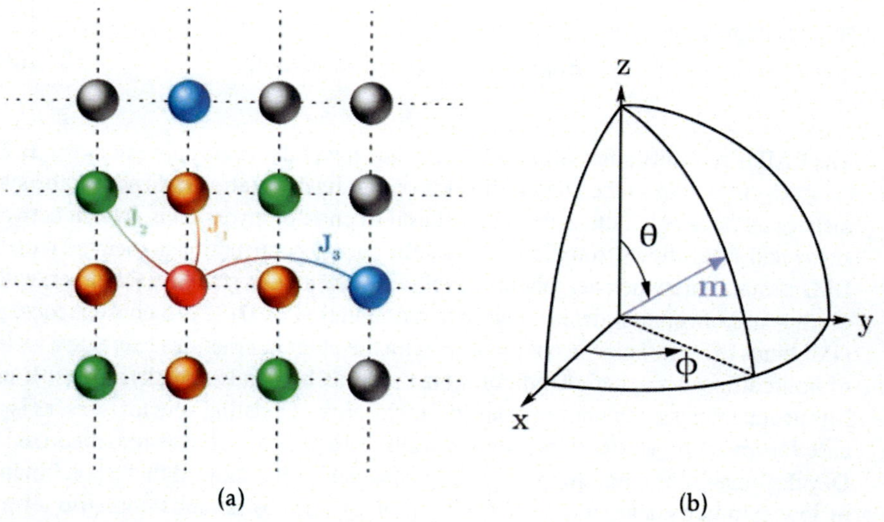

Fig. 2.1 **a** A simple cubic lattice, where J_1 is the isotropic exchange coupling between nearest neighbours, J_2 is between next-nearest neighbours, and J_3 is between next-next-nearest neighbours. Unless the exchange is frustrated, we restrict ourselves to nearest-neighbour interactions. **b** The spherical coordinate system used throughout this thesis

$$E_{\mathrm{ex}} = -\sum_{<ij>} J_{ij}\mathbf{m}_i \cdot \mathbf{m}_j, \tag{2.4}$$

in which the summation is done over pairs of neighbouring spins $< ij >$, with $i, j = 1 \ldots N$ and $i \neq j$. For the special case where J_{ij} is constant for all pairs $< ij >$, we write $J_{ij} = J_{\mathrm{ex}} = J_1$ if i and j are nearest neighbours, $J_{ij} = J_2$ if they are next-nearest neighbours, etc. (see Fig. 2.1a).

In the continuum limit, isotropic exchange may be expressed as

$$\mathscr{E}_{\mathrm{ex}} = \int \mathrm{d}V A\Big((\nabla m_x)^2 + (\nabla m_y)^2 + (\nabla m_z)^2\Big), \tag{2.5}$$

where $\nabla = \partial_x \mathbf{e}_x + \partial_y \mathbf{e}_y + \partial_z \mathbf{e}_z$ is the Nabla operator, $\mathrm{d}V$ is an elementary volume, and $A = J_{\mathrm{ex}} n/a$ is the exchange stiffness in units of energy per unit length (J/m), with n the number of atoms per unit cell. In all that follows, we restrict ourselves to the simple cubic lattice shown in Fig. 2.1a, for which $n = 1$.

2.1.1.2 The Dzyaloshinksii-Moriya Interaction (DMI)

The Dzyaloshinskii-Moriya interaction, or DMI, is an antisymmetric, anisotropic exchange term which arises in low symmetry crystals due to spin-orbit coupling (SOC). In the Heisenberg model, the total DMI energy of a system of N interacting spins is expressed as

$$E_{\mathrm{DM}} = -\sum_{<ij>} \mathbf{D}_{ij} \cdot \left(\mathbf{m}_i \times \mathbf{m}_j\right). \tag{2.6}$$

The DMI is typically a few orders of magnitude lower than isotropic exchange. In Eq. (2.6), $\mathbf{D}_{ij}$ is known as the Dzyaloshinskii vector. Its orientation depends on the symmetries of the crystal lattice and it vanishes if a centre of inversion is present between sites i and j. It follows from Eq. (2.6) that the energy contribution associated with the DMI is minimum when neighbouring spins are oriented normal to each other, with a given sense of rotation determined by the orientation of $\mathbf{D}_{ij}$. As a consequence, the DMI tends to rotate spins in competition with the other interactions to create a variety of non-collinear magnetic configurations, such as helicoids and skyrmions. It also lifts degeneracies by favouring a particular chirality of existing spin textures, as is the case for magnetic vortices and domain walls. This interaction was first theorized by Dzyaloshinskii [5] who, using symmetry considerations, showed that a combination of low symmetry and spin-orbit coupling gives rises to an antisymmetric term in the exchange. Moriya [6] then identified a microscopic mechanism which accounts for the presence of such a term in systems with spin-orbit coupling. DMI occurs in crystals with low symmetry such as non-centrosymmetric magnetic crystals—i.e., crystals lacking inversion symmetry. One of the first observations of its effect was weak ferromagnetism in antiferromagnetic compounds such as α-Fe_2O_3 and CrF_3, in which a net ferromagnetic moment emerges due to a small misalignment of the sublattice magnetization caused by antisymmetric exchange [6]. It was later found in chiral bulk magnets such as MnSi, in which a helical phase was observed at low temperature [7, 8].

DMI can also be induced in centrosymmetric crystals if the symmetry is broken by an external mechanism. This effect is weak in bulk materials, but becomes important in small magnetic structures such as thin films, multilayers, nanowires and nanodots. In thin films, the broken inversion symmetry at the surface is responsible for the emergence of interfacial DMI [9–12]. Strong DMI is these systems is engineered by combining a thin ferromagnetic film (e.g., Fe, Co) with a non-magnetic heavy metal substrate exhibiting strong spin-orbit coupling (Pt, Ir, etc.). The orientation of the $\mathbf{D}$-vector depends on the symmetries of the system. For a thin film lying in the xy-plane, such that the vertical inversion symmetry is broken along the z-axis, as is the case in Fig. 2.2a, the $\mathbf{D}$-vector is orthogonal to the displacement vector between atomic sites, and is expressed as $\mathbf{D}_{ij} = D\mathbf{e}_z \times \mathbf{r}_{ij}$, where $\mathbf{r}_{ij} = \mathbf{r}_j - \mathbf{r}_i$ is the in-plane displacement vector between sites i and j [11] (Fig. 2.2b). On the other hand, in bulk materials where the horizontal inversion symmetry is broken by the crystal lattice, the $\mathbf{D}$-vector is aligned with the displacement vector between atomic sites, and is expressed as $\mathbf{D}_{ij} = D\mathbf{r}_{ij}$ (Fig. 2.2c).

In the continuum approximation, the DMI is described by combinations of Lifshitz invariants [11],

$$L_{ij}^{(k)} = m_i \frac{\partial m_j}{\partial x_k} - m_j \frac{\partial m_i}{\partial x_k}, \tag{2.7}$$

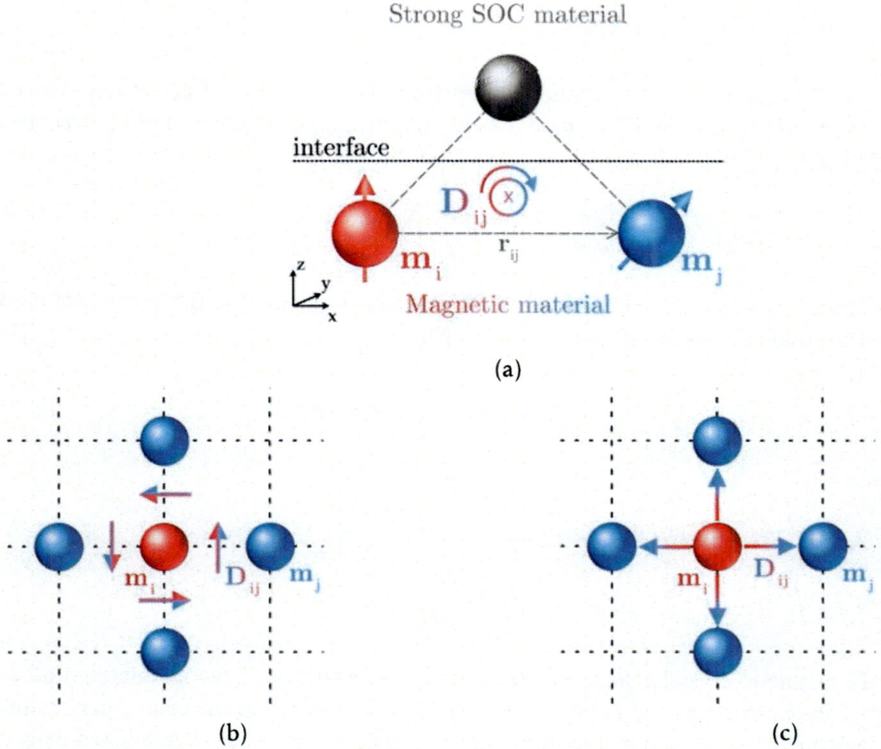

Fig. 2.2 **a** DMI induced by the vertical breaking of the inversion symmetry along z at an interface between a magnetic material and a material exhibiting strong spin-orbit coupling. **a** and **b** show different symmetries of the DMI in the xy-plane on a simple square lattice. **b** In the case where the DMI is induced at an interface via breaking of the inversion symmetry along z, the **D**-vector is orthogonal to the displacement vector between neighbouring sites. **c** In bulk materials with horizontal breaking of the inversion symmetry induced by the crystal lattice, the **D**-vector is collinear to the displacement vector between neighbouring sites

where i, j, k denote Cartesian indices. The interfacial form of the DMI with symmetry breaking along z can be written as

$$\mathscr{E}_{\mathrm{DM}} = \int \mathrm{d}V \mathscr{D}\Big(L_{zx}^{(x)} + L_{zy}^{(y)}\Big) = \int \mathrm{d}V \mathscr{D}\Big(\big[m_z \frac{\partial m_x}{\partial x} - m_x \frac{\partial m_z}{\partial x}\big] + id.(x \rightarrow y)\Big). \tag{2.8}$$

$\mathscr{D}$ is the continuum form of the DMI coupling constant expressed in units of energy per unit area (J/m^2), with $\mathscr{D} = 2D/a^2$ on the simple cubic lattice.

2.1.2 Zeeman Energy

The Zeeman energy contribution comes from the interaction of the system with an external magnetic field $\mathbf{B} = \mu_0 \mathbf{H}$, where μ_0 is the magnetic permeability of vacuum. It is expressed as

$$E_Z = -M_s \sum_{i=1}^{N} \mathbf{m}_i \cdot \mathbf{B}. \tag{2.9}$$

This term is minimized by aligning the magnetic moments along the external field. The continuous form of Eq. (2.9) is simply

$$\mathscr{E}_Z = -\mathscr{M}_s \int \mathrm{d}V \mathbf{m} \cdot \mathbf{B}. \tag{2.10}$$

2.1.3 Magnetic Anisotropy

2.1.3.1 Magnetocrystalline Anisotropy

In magnetic crystals, some crystallographic axes are more favourable than others for the magnetization to align along. This effect is known as the magnetocrystalline anisotropy. It originates from the interplay between spin-orbit coupling and magnetostatic interactions in the following way: the spins interact with the orbital motion by means of spin-orbit coupling, and the orbital motion in turns interacts with the crystal structure by means of the electrostatic fields and overlapping wavefunctions of neighbouring atoms [13]. We focus on systems in which the main magnetocrystalline anisotropy contribution gives rise to an easy axis, so that the associated energy takes on the form,

$$E_{\mathrm{an}} = -K_u \sum_{i=1}^{N} (\mathbf{m}_i \cdot \mathbf{e}_{\mathrm{an}})^2, \tag{2.11}$$

where $K_u > 0$ is the anisotropy constant, and $\mathbf{e}_{\mathrm{an}}$ is known as the easy axis. In the continuum limit,

$$\mathscr{E}_{\mathrm{an}} = -\mathscr{K}_u \int dV (\mathbf{m} \cdot \mathbf{e}_{\mathrm{an}})^2, \tag{2.12}$$

with $\mathscr{K}_u = K_u/a^3$ expressed in J/m^3 on the simple cubic lattice. In magnetic thin films lying in the xy-plane with a perpendicular easy axis such that $\mathbf{e}_{\mathrm{an}} = \mathbf{e}_z$, the interaction energy of $\mathbf{m}_i$ with the easy axis can simply be expressed as $-K_u m_{z,i}^2$.

2.1.3.2 Magnetostatic Contribution

Another contribution to the total anisotropy comes from dipole-dipole couplings between magnetic moments. Each magnetic moment constitutes a magnetic dipole and therefore contributes to the total dipolar field of the sample. Formally, the interaction potential that arises from the dipole-dipole coupling between two magnetic moments $\mathbf{m}_i$ and $\mathbf{m}_j$ takes the form

$$E_{ij,\mathrm{dip}} = \frac{I_{\mathrm{dip}}}{|\mathbf{r}_{ij}|^3}\Big(\mathbf{m}_i \cdot \mathbf{m}_j - 3(\mathbf{m}_i \cdot \hat{\mathbf{r}}_{ij})(\mathbf{m}_j \cdot \hat{\mathbf{r}}_{ij})\Big), \tag{2.13}$$

where $\mathbf{r}_{ij} = \mathbf{r}_j - \mathbf{r}_i$, $\hat{\mathbf{r}}_{ij} = \mathbf{r}_{ij}/|\mathbf{r}_{ij}|$, and I_{dip} is the strength of the dipolar coupling. The combined effect of the $N-1$ other magnetic moments on $\mathbf{m}_i$ constitutes the local dipolar field $\mathbf{H}_d^i = -\mu_0^{-1}\partial E_{\mathrm{dip}}/\partial \mathbf{M}_i$. In the continuum limit, the coupling of the magnetization with the dipolar—or demagnetizing—field $\mathbf{H}_d$ generated by the material constitutes the magnetostatic contribution to the energy. Combining $\mathbf{B} = \mu_0(\mathbf{H} + \mathcal{M})$ and Maxwell's equation, $\nabla \cdot \mathbf{B} = 0$, it follows that $\mathbf{H}_d$ is the field that arises from the divergence of the magnetization:

$$\nabla \cdot \mathbf{H}_d = -\nabla \cdot \mathcal{M}. \tag{2.14}$$

In ferromagnets, the interplay between magnetostatic energy and isotropic exchange is responsible for the emergence of uniformly magnetized domains with different orientations. This translates a tendency towards demagnetization, and is the reason why bulk magnets which have not been subjected to a magnetic field are usually found in a state of zero magnetization. $\mathbf{H}_d$ couples to the magnetization, $\mathcal{M}$, like an external field, in the form $-\frac{\mu_0}{2}\mathbf{H}_d \cdot \mathcal{M}$. In uniformly magnetized ferromagnets, the demagnetizing field may be expressed in terms of the demagnetizing tensor $\mathbf{N}$ as $\mathbf{H}_d = \mathbf{N} \cdot \mathcal{M}$.

In thin magnetic films, the pseudo magnetic charges on the surface are minimized if the magnetization lies in the plane of the film, so that the demagnetizing field can be approximated in the form $\mathbf{H}_d = -\mathcal{M}_z \mathbf{e}_z$ [14]. In the zero-thickness limit of ultrathin films, where dipolar couplings become local, this can be thought of as shape anisotropy, and the effective anisotropy constant becomes $\mathcal{K}_{\mathrm{eff}} = \mathcal{K}_u - \mu_0 \mathcal{M}_s^2/2$ [15]. $\mathcal{K}_{\mathrm{eff}} > 0$ defines an easy-axis and corresponds to perpendicularly magnetized thin films, while $\mathcal{K}_{\mathrm{eff}} < 0$ corresponds to an easy-plane configuration. In the rest of this work, we focus on the former case, and we write $\mathcal{K}_{\mathrm{eff}} = \mathcal{K}$, and $K = \mathcal{K} a^3$, to simplify the notation. In the end, in its atomistic form, the sum of the magnetocrystalline and magnetostatic contributions to the anisotropy in a thin magnetic film with perpendicular easy axis reduces to:

$$E_{\mathrm{an}} = -K \sum_{i=1}^{N} m_{z,i}^2, \tag{2.15}$$

and

$$\mathscr{E}_{\rm an} = -\mathscr{K} \int dV m_z^2. \tag{2.16}$$

2.1.4 Summary of Atomistic and Micromagnetics Quantities

In Table 2.1, we give a summary of the different physical quantities defined in this section with their corresponding symbol, unit, and the relations between the atomistic and micromagnetics quantities. We also define the following reduced dimensionless atomistic quantities that we shall use throughout this work: $j_\alpha = J_\alpha / J_1$ in which $(\alpha = 1, 2, 3)$; $d = |\mathbf{D}_{ij}|/J_1$; $k = K/J_1$; $b = BM_s/J_1$. In the previous relations, J_α corresponds to the isotropic exchange coupling constant between αth nearest neighbours, with $J_{\rm ex} = J_1$ for first nearest neighbours. The atomistic exchange coupling constants, J_α and D, are given per site so that, for instance, the total isotropic exchange energy between two parallel spins is $-2J_{\rm ex}$.

Table 2.1 Symbols, corresponding units and relation between atomistic and micromagnetics quantities. The atomistic exchange couplings are given per site

Quantity	Atomistic (simple cubic)			Micromagnetics		Relation
	Symbol	Unit	Reduced	Symbol	Unit	
Lattice constant	a	m	–	–		–
Saturation magnetization	M_s	A m^2	-	$\mathscr{M}_s$	A/m	$\mathscr{M}_s = M_s/a^3$
Isotropic exchange strength	$J_{\rm ex} = J_1$	J	1	A	J/m	$A = J_1/a$
DMI strength	D	J	$d = D/J_1$	$\mathscr{D}$	J/m^2	$\mathscr{D} = 2D/a^2$
Anisotropy constant	K	J	$k = K/J_1$	$\mathscr{K}$	J/m^3	$\mathscr{K} = K/a^3$
Magnetic field (induction)	B	T	$b = BM_s/J_1$	Idem atomistic		–
Magnetic permeability of vacuum	μ_0	N/A^2	–	Idem atomistic		–

2.2 Ferromagnetic Domain Walls

The state of a ferromagnetic crystal which has not been subjected to an external magnetic field is usually that of zero magnetization. This is explained by the fact that crystals are divided into a finite number of regions of uniform magnetization, known as magnetic domains, such that, in the demagnetized state, the resulting magnetization of the crystal is zero. The domain boundaries, or domain walls, are regions usually much smaller than the domains themselves, in which the magnetization gradually rotates from one orientation to the other (Fig. 2.3a and b). They pertain to a class of solutions known as (static) topological solitons, i.e., solitonic solutions that are statically stable due to topological constraints. The formation of domains lowers the magnetostatic energy of the sample by reducing the magnetic charges on the surface, but raises the exchange energy by creating regions where the spins are not collinear. For that reason, domains are favourable if the size of the system is big enough as to ensure that the reduction in magnetostatic energy is higher than the gain in exchange energy.

2.2.1 Characteristic Wall Profile

The width of a domain wall results from the competition between anisotropy and isotropic exchange energies. The more gradual the rotation of the magnetization, the lower the exchange energy. On the other hand, the anisotropy favours particular

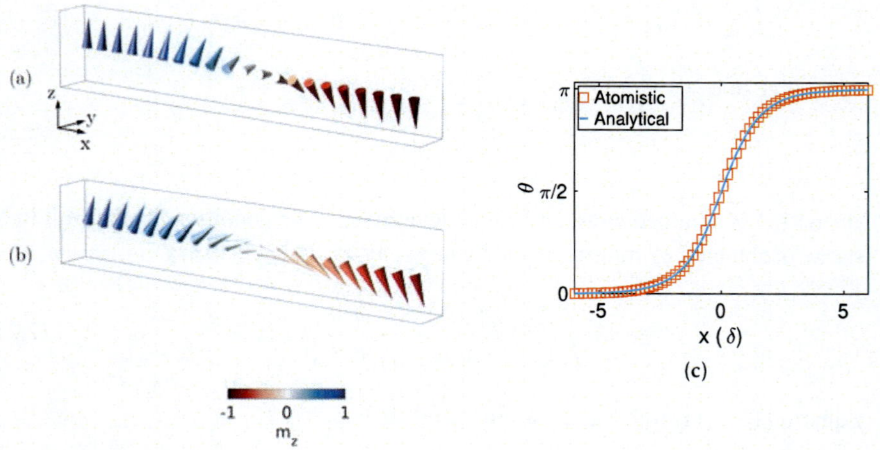

Fig. 2.3 **a** Bloch π-wall in which the magnetization rotates in the plane of the wall, **b** Néel π-wall in which the magnetization rotates perpendicular to the plane of the wall, **c** analytical micromagnetic and numerical atomistic π-domain wall profiles as a function of the spatial displacement from the wall centre expressed in units of the characteristic wall width, δ (Eq. (2.31)). The micromagnetic analytical expression of $\theta(x)$ is given in Eq. (2.32), where θ is the angle of the magnetization with the easy-axis. Here, it corresponds to the polar angle with the z-axis, as defined in Fig. 2.1b and Eq. (2.3)

orientations of the magnetization along the easy axes and/or easy planes of the material. In what follows, we derive the characteristic quantities in a one-dimensional wall—i.e., wall profile, wall width, and wall energy. The derivation is based on that of Ref. [1].

We consider a one-dimensional chain of magnetic moments along the x-axis, with an easy-axis along the z direction. We look at a π-wall separating two domains in which the magnetization goes from $+z$ to $-z$, as shown in Fig. 2.3. $\theta = \cos^{-1} m_z$ is the angle between $\mathbf{m}$ and the $z-$axis (Fig. 2.1b), and we assume that the rotation occurs in a great enough number of steps for $\theta(x)$ to be considered a continuous variable. In this context, the interactions can then be written in the continuum approximation. If $f(\theta)$ is the anisotropy energy per unit volume, the total energy density per unit area of the wall is given by the sum of the exchange energy contribution, σ_{ex}, and the anisotropy energy contribution, σ_{an},

$$\sigma = \sigma_{\text{ex}} + \sigma_{\text{an}} = \int_{-\infty}^{+\infty} \left[A\left(\frac{\partial \theta}{\partial x}\right)^2 + f(\theta) \right] dx. \tag{2.17}$$

A small change, $\delta\theta$, in θ induces the following change in σ:

$$\delta\sigma = \int_{-\infty}^{+\infty} \left[2A \frac{\partial \theta}{\partial x}\frac{\partial}{\partial x}\delta\theta + \frac{\partial f(\theta)}{\partial \theta}\delta\theta \right] dx. \tag{2.18}$$

We integrate the first term by parts:

$$\int_{-\infty}^{+\infty} 2A \frac{\partial \theta}{\partial x}\frac{\partial}{\partial x}\delta\theta = \left[2A\frac{\partial \theta}{\partial x}\delta\theta \right]_{-\infty}^{+\infty} - \int_{-\infty}^{+\infty} 2A\frac{\partial^2 \theta}{\partial x^2}\delta\theta dy, \tag{2.19}$$

and notice that

$$\left[2A\frac{\partial \theta}{\partial x}\delta\theta \right]_{-\infty}^{+\infty} = 0, \tag{2.20}$$

since $\theta(x)$ is assumed constant inside the domains. The condition for the wall to be stable is obtained by minimizing the energy: $\delta\sigma = 0$, that is to say,

$$\delta\sigma = \int_{-\infty}^{+\infty} \left[\frac{\partial f(\theta)}{\partial \theta} - 2A\frac{\partial^2 \theta}{\partial x^2} \right] \delta\theta dx = 0. \tag{2.21}$$

Equation (2.21) is true for any arbitrary $\delta\theta$ if

$$\frac{\partial f(\theta)}{\partial \theta} - 2A\frac{\partial^2 \theta}{\partial x^2} = 0, \tag{2.22}$$

which is the Euler equation of the one-dimensional variational problem. Multiplying Eq. (2.22) by $\partial\theta/\partial x$ and integrating, we obtain an expression for the anisotropy energy:

$$\int \frac{\partial f(\theta)}{\partial \theta}\frac{\partial \theta}{\partial x}dx = 2A\int \frac{\partial^2 \theta}{\partial x^2}\frac{\partial \theta}{\partial x}dx, \tag{2.23}$$

$$f(\theta) = A\left(\frac{\partial \theta}{\partial x}\right)^2, \tag{2.24}$$

which is the same as the exchange energy, and from Eq. (2.24),

$$dx = \sqrt{A}\frac{d\theta}{\sqrt{f(\theta)}} \tag{2.25}$$

The energy per unit area of the wall is therefore

$$\begin{aligned} \sigma &= 2\int f(\theta)dx, \\ &= 2\int f(\theta)\frac{\partial x}{\partial \theta}d\theta, \\ &= 2\sqrt{A}\int \sqrt{f(\theta)}d\theta, \end{aligned} \tag{2.26}$$

where we have used Eq. (2.25) in the last step. With an uniaxial anisotropy, $f(\theta) = \mathscr{K}\sin^2\theta$, and so the energy of a π-wall is

$$\sigma_0 = 2\sqrt{A\mathscr{K}}\int_0^\pi \sin\theta d\theta = 4\sqrt{\mathscr{K}A}. \tag{2.27}$$

We note that with the above choice of $f(\theta)$, Eq. (2.22) becomes

$$\sin\theta - \frac{A}{\mathscr{K}}\frac{\partial^2\theta}{\partial x^2} = 0, \tag{2.28}$$

which is the static sine-Gordon equation (Eq. (1.1)). By choosing the origin at the centre of the wall, it follows from Eq. (2.25) that the wall profile is expressed as

$$x(\theta) = \sqrt{\frac{A}{\mathscr{K}}}\int_{\pi/2}^{\theta}\frac{d\theta'}{\sin\theta'} = \sqrt{\frac{A}{\mathscr{K}}}\ln\left(\tan\frac{\theta}{2}\right). \tag{2.29}$$

The characteristic (asymptotic) width of the wall δ emerges from Eq. (2.29),

$$\exp\left(\frac{x}{\sqrt{A/\mathscr{K}}}\right) = \exp\left(\frac{x}{\delta}\right) = \tan\frac{\theta}{2}, \tag{2.30}$$

where

$$\delta = \sqrt{A/\mathcal{K}}, \tag{2.31}$$

and

$$\theta(x) = 2\arctan\left(e^{x/\delta}\right), \tag{2.32}$$

which we plot as a continuous blue line in Fig. 2.3c.

We distinguish between Bloch walls [16], in which the magnetization rotates in the plane of the wall (Fig. 2.3a), and Néel walls, in which the magnetization rotates perpendicular to the wall plane (Fig. 2.3b). Depending on the sample's shape and the magnetization direction inside domains, the type of wall that minimizes the magnetostatic energy is favoured against the other. In thin films magnetized out-of-plane, and in the absence of DMI, Bloch walls are more favourable as they minimize volume charges—i.e., the magnetic charge density $\rho_m = -\nabla \cdot \mathbf{m}$ reduces to $\rho_m = -dm_x/dx = 0$ in the Bloch wall, where x is the propagation direction. This result, however, becomes less true in ultrathin films, since the energy difference between Bloch and Néel walls decreases with the film thickness [17]. Additionally, in the absence of chiral couplings, clockwise and counterclockwise directions of the magnetization rotation inside the wall are degenerate.

2.2.2 *Effect of Chiral Couplings*

In thin films with perpendicular anisotropy, the competition between the DMI and magnetostatic energy results in a progressive transition from Bloch walls at $\mathcal{D} = 0$, to Néel walls with increasing $\mathcal{D}$. The DMI also selects a preferred handedness, or chirality, of the wall. The wall energy in the presence of DMI becomes $\sigma = \sigma_0 \pm \pi\mathcal{D}$ [18, 19]. When the chirality of the wall is the one favoured by the DMI, $\sigma = \sigma_0 - \pi\mathcal{D}$ and the wall is more energetically favourable than its counterpart of opposite chirality. Domain walls in PMA materials under uncompensated DMI have been named Dzyaloshinskii walls [19]. Above a certain critical value $\mathcal{D}_c$ of the DMI, expressed as $\mathcal{D}_c = 4\mathcal{K}\delta/\pi = \sigma_0/\pi$, the spin spiral state replaces the collinear state as the ground state of the system.

2.2.3 *Winding Number and Topology*

We now consider two π-walls, separating three magnetic domains. In the previously considered planar spin model, the spins are arranged along a one-dimensional chain and rotate strictly into the easy plane, such that $\theta(x)$ describes the magnetization. The order-parameter space is the unit circle, S^1. In this configuration where $\theta(x \to \pm\infty) = 0$, the one-dimensional physical space, $\mathbb{R}^1$, is topologically equivalent to S^1, and we have the map $f : S^1 \to S^1$, which may be classified by the homotopy class $\pi_1(S_1) = \mathbb{Z}$. Topologically distinct configurations are then defined by how many

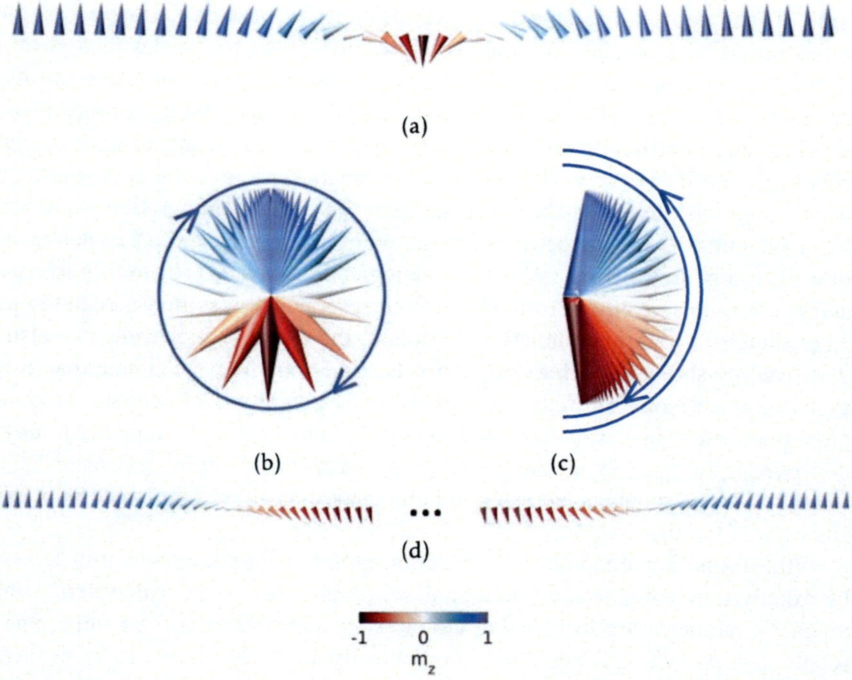

Fig. 2.4 **a**, **b** Pair of Néel π-walls under a perpendicular applied field with fixed sense of rotation where **b** shows the stereographic projection onto the unit circle, S^1. The magnetization rotates in two-dimensional space from $\theta = 0$, to π, then from π to 2π, which results in a winding number $n = 1$, and thus a topologically nontrivial spin texture with respect to the easy plane (soliton-soliton pair). In this case, the domain of down-magnetization in the middle can be reduced in size by a magnetic field applied along +z. **c**, **d** Pair of Néel π-walls at zero applied field with two different senses of rotation where **c** shows the stereographic projection onto the unit circle, S^1. The magnetization rotates in two-dimensional space from $\theta = 0$, to π, then from π to 0, which yields a vanishing winding number, $n = 0$, associated with a topologically trivial state (soliton-antisoliton pair). In this case, applying a magnetic field along +z tends to destroy the walls and results in a single uniformly magnetized state along the field direction

times the magnetization wraps the unit circle S^1. This is measured by the winding number—or degree of mapping—deg $f = n$, defined as [20]

$$n = \frac{1}{2\pi} \int_{-\infty}^{+\infty} dx \partial_x \theta. \tag{2.33}$$

If the two π-walls exhibit the same sense of rotation—or chirality—as is the case in Fig. 2.4a, the magnetization rotates from $\theta = 0$ to π, and then from π to 2π, and therefore wraps around the unit circle one time. This can be visualized through the stereographic projection of the chiral wall shown in Fig. 2.4b. In this case, the winding number is $n = \pm 1$, and the structure is topologically nontrivial with respect to the easy plane. This configuration is referred to as soliton-soliton pair. In the continuum limit, it is topologically stable since, within the strict planar rotation model, there exists no continuous path in parameter-space that leads to the destruction of the wall. In particular, adding a DMI contribution to the present structure results in the selection of a specific sense of rotation—or chirality—which in turn favours topologically nontrivial pair of walls. On the other hand, in the absence of chiral couplings and if both π-walls possess an opposite sense of rotation, θ goes from 0 to π, and back to 0, which in turn yields $n = 0$. This configuration is shown in Fig. 2.4d, and c shows the corresponding stereographic projection. The achiral pair of walls is topologically trivial, as it is topologically equivalent to the collinear state, and is not robust against deformations. It corresponds to a soliton-antisoliton pair. Nevertheless, within an atomistic description, the chiral pair of walls can also be destroyed by strong enough external effects. Additionally, we can note that in real systems, anisotropies are finite, and so the strict planar model does not hold. The order parameter space is, in fact, the 2-sphere S^2, and the pair of chiral walls may be destroyed if the spins rotate along y (case of "escape via the third dimension" [21]). Indeed, in a one-dimensional chain, topologically nontrivial solutions do not exist, since $\pi_1(S^2) = 0$.

While the notion of topological protection applies to the continuum limit, it cannot be exactly transposed to a discretized description of space. In real systems where magnetic moments are localized at atomic sites, energy barriers are finite, and so topological protection is not strict. Topologically nontrivial states can be destroyed if the energy barrier than separates them from the uniform state is overcome by external effects and/or thermal fluctuations.

2.3 Magnetic Skyrmions

2.3.1 General Characteristics

Magnetic skyrmions are topologically nontrivial, particle-like, solitonic magnetic textures which are exponentially localized in space. They typically exist in two dimensions, but can extend to three dimensions as skyrmion strings. Figure 2.5 shows spin maps of some typical skyrmion configurations which we will describe later in this section. Originally, skyrmions were solitonic solutions derived by Skyrme [22] within his model for low energy dynamics between mesons and baryons. Skyrmions from the initial Skyrme model live on the three-dimensional space, and the term "baby skyrmions" was later introduced as solutions of the two-dimensional so-called "baby Skyrme" model [23, 24]. In that sense, magnetic skyrmions are similar to baby skyrmions. Within a skyrmion, the spins smoothly rotate in three dimensional space and their norm remains fixed, such that the order-parameter space is now the unit sphere, S^2. For an isolated skyrmion configuration, we assume that the magnetization tends asymptotically to a uniform value at infinity, i.e., $\mathbf{m}(\rho \to \infty) = \text{cst}$. The physical space $\mathbb{R}^2$ may then be stereographically projected onto the punctured unit

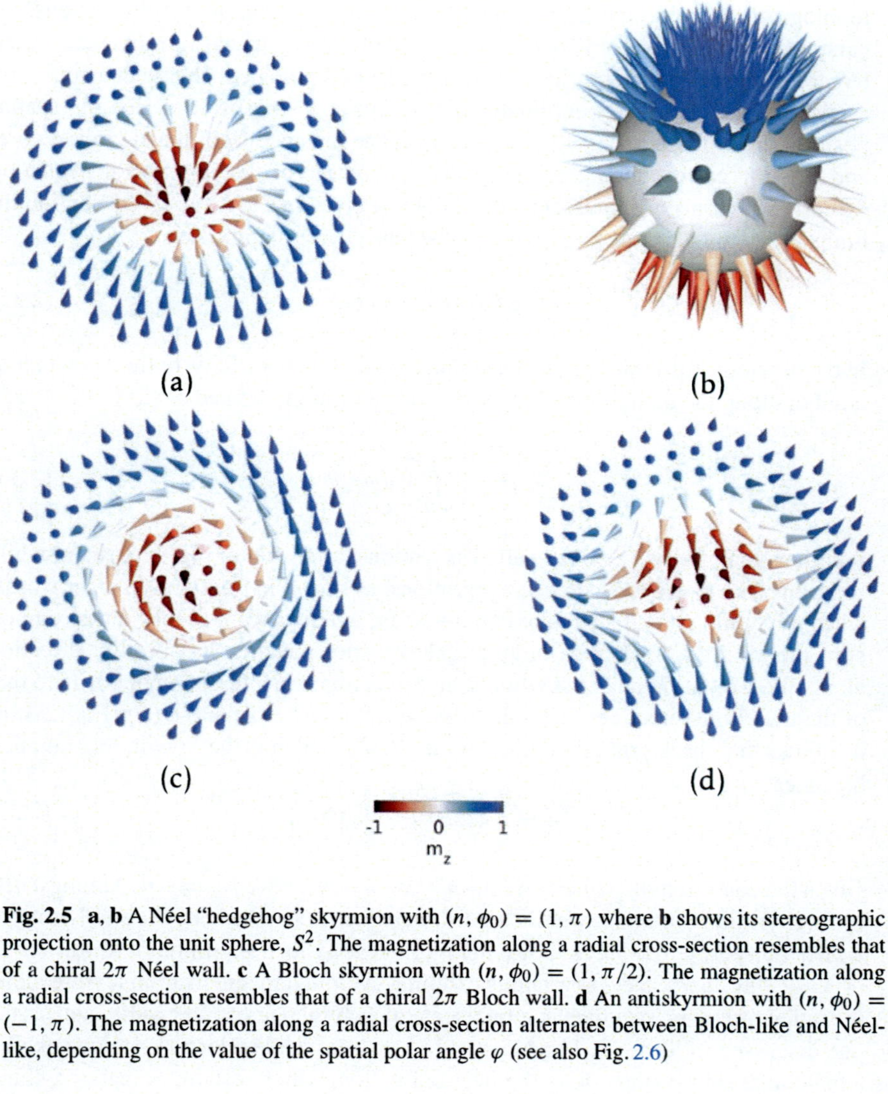

Fig. 2.5 **a, b** A Néel "hedgehog" skyrmion with $(n, \phi_0) = (1, \pi)$ where **b** shows its stereographic projection onto the unit sphere, S^2. The magnetization along a radial cross-section resembles that of a chiral 2π Néel wall. **c** A Bloch skyrmion with $(n, \phi_0) = (1, \pi/2)$. The magnetization along a radial cross-section resembles that of a chiral 2π Bloch wall. **d** An antiskyrmion with $(n, \phi_0) = (-1, \pi)$. The magnetization along a radial cross-section alternates between Bloch-like and Néel-like, depending on the value of the spatial polar angle φ (see also Fig. 2.6)

sphere S^2 with the missing point corresponding to $\rho \to \infty$, and we have the map $f : S^2 \to S^2$, which may be classified by the homotopy class $\pi_2(S^2) = \mathbb{Z}$. [20, 25]. In Fig. 2.5b, we show the stereographic projection of the Néel skyrmion, as shown in Fig. 2.5a, onto S^2. We define the skyrmion number, or topological charge, [26],

$$N_s = \int \mathbf{dr}^2 \rho_s = \frac{1}{4\pi} \int \mathbf{dr}^2 \mathbf{m} \cdot (\partial_x \mathbf{m} \times \partial_y \mathbf{m}), \tag{2.34}$$

which corresponds to the degree of mapping, $N_s = \deg f$, and counts the number of times the spin configuration wraps the unit sphere, S^2. Skyrmions carry an integer topological charge, which implies that they are robust against small deformations and cannot be continuously unwound into the topologically trivial ferromagnetic state ($N_s = 0$). ρ_s, as defined by Eq. (2.34), is the local topological charge density.

We can use the polar coordinates $\rho = (\rho\cos\varphi, \rho\sin\varphi)$ to describe the spatial position, in which $\rho = \sqrt{x^2 + x^2}$ is the in-plane distance from the skyrmion core, and φ is the corresponding azimuth. The orientation of the spin vector $\mathbf{m}$ can be expressed in terms of spherical angles (θ, ϕ) as given in Eq. (2.3). In the continuum limit, $\theta(\rho)$ only depends on ρ, and the in-plane skyrmion profile is

$$\phi(\varphi) = n\varphi + \phi_0. \tag{2.35}$$

In the previous expression, ϕ_0 is the helicity, and n is the vorticity, or the S^1- winding number along the domain wall delimiting the skyrmion, defined as [25]

$$n = \frac{1}{2\pi}\int_0^{2\pi} d\varphi\partial_\varphi\phi. \tag{2.36}$$

Solutions that verify $n > 0$ are called skyrmions, as shown in Fig. 2.5a and c, while solutions with $n < 0$ are called antiskyrmions, as shown in Fig. 2.5d. In other words, when travelling around a circular contour at a fixed distance ρ to the (anti)skyrmion core, the winding direction of the spins in a skyrmion matches the travelling direction along the contour. In an antiskyrmion, the spins rotate in a direction opposite to that of the travelling direction. From the symmetric ansatz $\theta(\rho)$ and $\phi(\varphi)$ (Eq. (2.35)), it follows that the skyrmion number in Eq. (2.34), which is the S^2-winding number, becomes

$$N_s = \frac{n}{2}\Big(m_z(0) - m_z(\infty)\Big). \tag{2.37}$$

Given the conditions $m_z(0) = -1$ and $m_z(\infty) = 1$, we have $N_s = -n$. Amongst first order skyrmions ($n = 1$), we distinguish between Néel—or hedgehog—skyrmions with helicity $\phi_0 = 0$ or π, as shown in Fig. 2.5a, and Bloch skyrmions with $\phi_0 = \pi/2$ or $3\pi/2$ (Fig. 2.5c). Néel skyrmions exhibit a cycloidal magnetization texture along the radial direction, ρ, which is reminiscent of a chiral pair of Néel walls, such as the one discussed in Sect. 2.2.3 and shown in Fig. 2.4a, while Bloch skyrmions exhibit a helicoidal spin texture along the radial direction, which resembles that of a chiral pair of Bloch walls. In antiskyrmions ($n = -1$), on the other hand, the spin texture along the radial direction alternates between a Néel-type and a Bloch-type pair of walls depending on the spatial angle φ. We show the detail of the spin texture of an antiskyrmion on the simple square lattice in Fig. 2.6, in which the radial directions corresponding to lattice axes x and y ($\varphi = \{0, \pi/2\}$) display clockwise- (CW) and counter-clockwise- (CCW) rotating Néel-type profiles, while along the lattice diagonals ($\varphi = \{\pi/4, 3\pi/4\}$), the magnetization is that of CW and CCW Bloch-type profiles. In higher-order skyrmion solutions ($|n| > 1$), in which the magnetization wraps the unit sphere more than once, the radial profile also alternates between Néel and Bloch. For instance, typical spin textures corresponding to a second-order

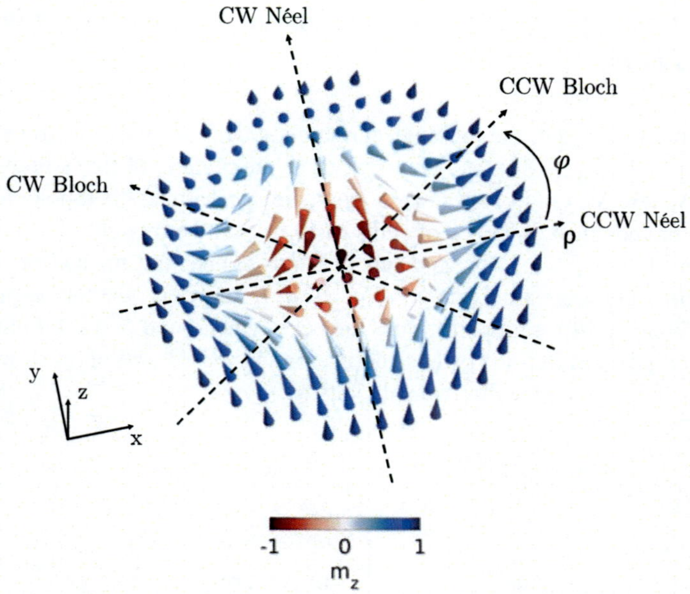

Fig. 2.6 Anatomy of the antiskyrmion spin texture. The winding direction of the spins in the direction of increasing φ is negative (i.e., clockwise (CW) rotation of the magnetization for a counter-clockwise (CCW) rotation along the contour), yielding a negative winding number, $n = -1$. Depending on the value of φ, the spin texture along the radial direction ρ alternates between Néel-type and Bloch-type with CW and CCW rotation of the spins

skyrmion ($n = 2$), and a second-order antiskyrmion ($n = -2$) are shown respectively in Fig. 2.9d and e.

Magnetic skyrmions are stabilized in ferromagnets by the introduction of a characteristic lengthscale via competing interactions. This is typically achieved in non-centrosymmetric magnets, in which chiral couplings arise in the form of DMI. That particular type of solution, commonly referred to as chiral skyrmions, were theoretically predicted and investigated in the 1990s [27–29]. In magnets with inversion symmetry, skyrmions can be stabilized via dipolar interactions (skyrmion bubbles, also referred to as bubble domains) [30–35], as well as frustrated exchange couplings [36–41]. In the rest of this section, we firstly define the topological charge on the discrete spin lattice. After that, we give details on the different categories of magnetic skyrmions according to their stabilization mechanism.

2.3.2 *Topological Charge Definition on the Atomistic Spin Lattice*

In a continuum model, the topological charge is defined as its integral expression that we gave in Eq. (2.34), and corresponds to the number of times $\mathbf{m}(x, y)$ wraps the unit sphere S^2. On the discretized atomistic spin lattice, the topological charge needs to be redefined. A lattice of spins $\mathbf{m}(x_i, y_i)$ can be thought of as a net on S^2, with vertices $\mathbf{m}(x_i, y_i) = \mathbf{m}_i$, and rubber bands connecting $\mathbf{m}_i$ and $\mathbf{m}_j$ whenever sites i and j are nearest neighbours (Fig. 2.7a). In that case, the topological charge can be defined as the number of times the net winds around S^2 [43]. Following the expression given in Ref. [44] whose derivation was based on Ref. [43], we give the topological charge on the discrete spin lattice,

$$N_s = \frac{1}{4\pi} \sum_l q_l, \tag{2.38}$$

with

$$\tan\left(\frac{q_l}{2}\right) = \frac{[\mathbf{m}_i, \mathbf{m}_j, \mathbf{m}_k]}{1 + \mathbf{m}_i \cdot \mathbf{m}_j + \mathbf{m}_i \cdot \mathbf{m}_k + \mathbf{m}_j \cdot \mathbf{m}_k}. \tag{2.39}$$

The notation $[\mathbf{m}_i, \mathbf{m}_j, \mathbf{m}_k] = \mathbf{m}_i \cdot (\mathbf{m}_j \times \mathbf{m}_k)$ corresponds to the scalar triple product. The index l runs over all the elementary triangles of the lattice, where $\mathbf{m}_i$, $\mathbf{m}_j$, $\mathbf{m}_k$ define the vertices of a triangle (Fig. 2.7b). The indices i, j, k are attributed by rotating counterclockwise about the z-axis.

$$Q_l = \frac{q_l}{4\pi} \tag{2.40}$$

is a discrete equivalent to the topological charge density ρ_s (Eq. (2.34)) and may be interpreted as a local topological charge. The topological charge on the discrete spin lattice defined through Eqs. (2.38) and (2.39) should hold for all types of lattices.

In Fig. 2.7a, we show a "ball in a net" representation of the Néel skyrmion in Fig. 2.5a, which was also somewhat inspired by Fig. 3 in Ref. [45]. Vertices correspond to the tip of the magnetic vectors with their origin in the centre of the sphere, and the edges represent the exchange coupling between first neighbours. The solid sphere inside the net represents the unit sphere. The ferromagnetic background (spins along $+z$) is found at the north pole of the sphere, while in this thesis, the core of a skyrmion typically points along the south pole (along $-z$). The view is set just below the south pole, looking axially towards the $+z$ direction, so that the spin in the centre of the view corresponds to the core of the skyrmion. Annihilating a nontrivial topological texture to recover the uniformly magnetized state involves bringing the mesh back to the north pole by moving the vertices on the sphere. This can be thought of as taking a ball (solid sphere) out of a net (dark blue mesh) by deforming the rubber band net.

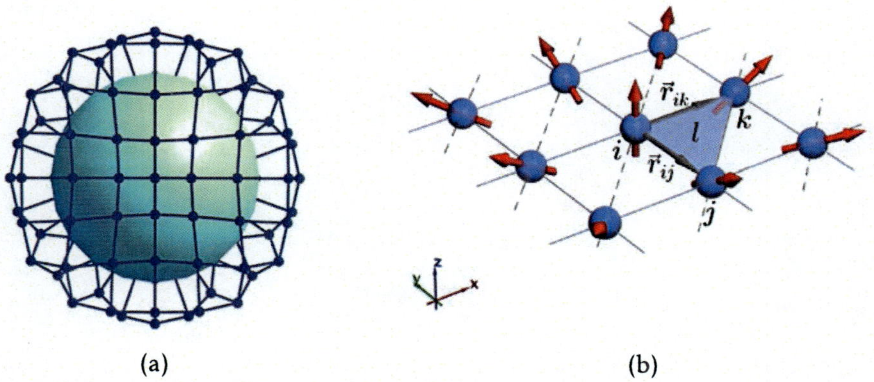

Fig. 2.7 **a** Mapping of the spin configuration for an isolated skyrmion onto the unit sphere. Vertices correspond to the tip of the magnetic vectors with their origin in the centre of the sphere, and the edges represent the exchange coupling between first neighbours. The solid sphere inside the net represents the unit sphere. **b** Elementary triangles on the simple square lattice (image source: [42])

2.3.3 Chiral Skyrmions in Non-centrosymmetric Magnets

The first theoretical prediction that the DMI can stabilize two-dimensional solitons dates back to 1989 [27]. Initially labelled "(chiral) magnetic vortices", they were later renamed "chiral magnetic skyrmions" to mark the distinction with vortices, in which the magnetization only spans over half of the unit sphere, yielding a half-integer topological charge (see Fig. 1.1b). A skyrmion lattice was first detected in the bulk of the B20 chiral magnet MnSi in reciprocal space, via neutron scattering [46]. Shortly after, Yu et al. reported the first real space observation of a skyrmion lattice in a thin film of $Fe_{0.5}Co_{0.5}$ by means of Lorentz transmission electron microscopy (TEM) [47]. These observations opened the way of potential spintronics applications of skyrmions in data storage and logic devices [48–52]. The existence of skyrmions was then confirmed in many other types of materials [53–56]. In particular, some well-studied transition-metal thin films and multilayers that give rise to a substantial interfacial DMI in which individual skyrmions were experimentally observed are Ir/Co/Pt [57], Pt/Co/AlO_x [58, 59], and Pd/Fe/Ir(111) [53, 60–63]. In chiral magnets, in the absence of any other effect, the competition between isotropic exchange and DMI results in a ground state of spatially modulated spin spirals (SSs) with a fixed sense of rotation. Additional couplings such as anisotropies and external field can suppress the SS state in favour of the uniformly magnetized ferromagnetic (FM) state, from which isolated skyrmions can be stabilized as (meta)stable excitations. In the presence of perpendicular applied magnetic field, a thermodynamically stable skyrmion lattice phase appears at intermediate field values [29, 64].

The symmetries of the DMI favour a single class of solutions with fixed chirality and helicity. For instance, interfacial DMI favours Néel skyrmions (Fig. 2.5a), and bulk DMI favours Bloch skyrmions (Fig. 2.5c). Antiskyrmions (Figs. 2.5d and 2.6)

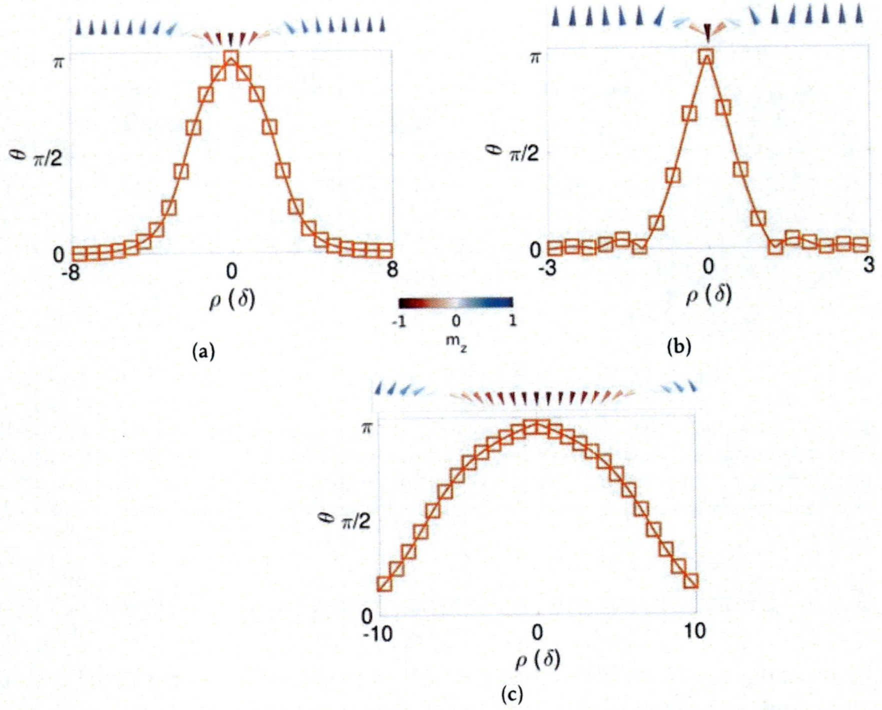

Fig. 2.8 Numerical profile obtained from atomistic simulations (squares) and corresponding spline interpolation (line) for skyrmions solutions with different stabilization mechanisms, as a function of the radial spatial displacement from the skyrmion core, ρ, expressed in units of the characteristic domain wall width, δ (Eq. 2.31). θ is the polar angle with the easy axis, z. The spin maps at the top show the spin texture along a radial cross-section of the skyrmion. **a** Profile of a chiral Néel skyrmion stabilized by the DMI. The core is strongly localized, with linear variations of the polar angle, while the asymptotic profile of the tail far from the centre resembles that of a ferromagnetic domain wall. **b** Profile of a Néel skyrmion stabilized by frustrated exchange. The core is strongly localized. In the tail, the polar angle oscillates as a function of distance, indicating a reversal of helicity. **c** Profile of a Néel skyrmion bubble stabilized by dipolar couplings. The core is spatially extended and not strongly localized

can be stabilized by anisotropic DMI [65, 66]. This has for instance been reported in tetragonal Heusler compounds, in which antiskyrmions were observed via Lorentz TEM [55]. Replacing $\mathbf{D}$ with $-\mathbf{D}$ results in skyrmion solutions of the opposite chirality. Nonetheless, the possibility to stabilize structures with an arbitrary integer topological charge in chiral magnets ("skyrmionic sacks") has recently been explored [67]. Far from the skyrmion core, the asymptotic behaviour of the spins along the radial direction ρ is that of a ferromagnetic wall, $\theta(\rho) \sim \kappa e^{\pm\rho/\delta}$, in which δ is the characteristic domain wall width (Eq. (2.31)) and $\kappa = \pi\mathscr{D}/4\sqrt{A\mathscr{K}}$, whereas the core is strongly localized, arrow-like shaped, with linear variation of the polar angle:

$\theta(\rho) \sim \rho\mathscr{D}/A$ [68, 69]. We show a typical chiral skyrmion profile along a radial cross-section in Fig. 2.8a, where the spin texture is shown at the top. It follows that the characteristic size of chiral skyrmions is similar to that of ferromagnetic walls, while being spatially localized in all directions. Besides δ, the other characteristic lengthscale in the system is $\zeta = 2A/\mathscr{D}$ [35, 59], which characterizes the period of the spatially modulated states that arise from the competition of isotropic exchange and DMI. There is no exact analytical expression for skyrmion profiles, however the double soliton ansatz of two overlapping π-walls at positions $\pm c$ works well [21, 60, 70]:

$$\theta(\rho, c, \delta) = \sum_{+,-} \arcsin\left(\tanh\left(\frac{-\rho \pm c}{\delta}\right)\right), \tag{2.41}$$

in which the sum is carried out over the two π-wall profiles with spatial displacement given respectively by $\rho + c$ and $\rho - c$.

2.3.4 Skyrmions in Frustrated Magnets

Frustrated isotropic exchange corresponds to the case where the coupling constant J_α changes sign depending on the shell of neighbours $\alpha = 1, 2, 3\ldots$, resulting in alternating ferromagnetic ($J_\alpha > 0$) and antiferromagnetic ($J_\alpha < 0$) couplings. For clarity, $\alpha = 1$ corresponds to nearest neighbours, $\alpha = 2$ to next-nearest neighbours, etc. (see Fig. 2.1a). Exchange frustration has been shown to arise from interface effects in certain systems. Pd/Fe/Ir(111) is an experimentally well-studied system that was reported to host interface nanoscale skyrmions [53, 60]. Density functional theory (DFT) calculations have shown that for fcc stacking of the Pd layer, the isotropic exchange is ferromagnetic between first neighbours, but antiferromagnetic beyond first neighbours, leading to a SS ground state even without the inclusion of SOC (i.e., DMI) in the calculations [61]. More recently, a strong exchange frustration was reported in Rh/Fe/Ir(111), which was found to be a driving force behind the SS ground state [71]. Strong exchange frustration was also very recently revealed in Rh/Co/Ir(111) by spin-polarized scanning tunneling microscopy (SP-STM) measurements combined with DFT calculations [72]. In this system, skyrmions of approximately 5 nm in diameter were observed at 4K at zero magnetic field in the virgin FM state. The exchange frustration was found to play an important role in their stabilization. For these reasons, studying skyrmion solutions stabilized by frustrated exchange is important for technological applications. In exchange-frustrated magnets, spatially modulated states can arise without the need for chiral couplings. For certain specific values of the exchange parameters, a spin-spiral ground state is realized in the absence of other interactions [36–41]. Similar to the case of chiral magnets, the SS state is suppressed by additional couplings such as external field and/or magnetic anisotropies, and the uniformly magnetized ferromagnetic state is recovered via a second-order phase transition [39]. From that state, metastable skyrmion solu-

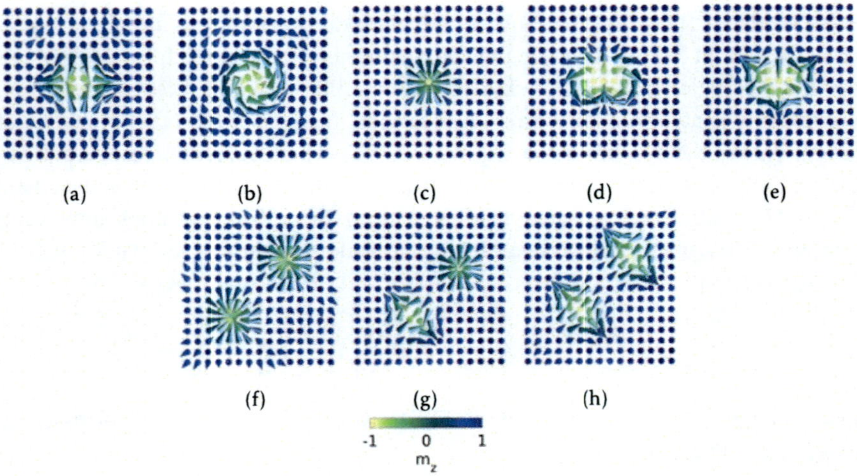

Fig. 2.9 Different skyrmion solutions stabilized in a frustrated magnet with inversion symmetry: **a** antiskyrmion with $(n, \phi_0) = (-1, 0)$, **b** skyrmion with $(n, \phi_0) = (1, \pi/2)$ (Bloch), **c** skyrmion with $(n, \phi_0) = (1, \pi)$ (Néel), **d** 2nd order skyrmion with $(n, \phi_0) = (2, 3\pi/2)$, **e** 2nd order antiskyrmion with $(n, \phi_0) = (-2, \pi/2)$, **f** skyrmion pair, **g** skyrmion/antiskyrmion pair, **h** antiskyrmion pair

tions can be stabilized. In the presence of a perpendicular easy-axis anisotropy, the skyrmion lattice becomes thermodynamically stable at an intermediate applied field [39].

In comparison to chiral skyrmions stabilized by the DMI, where the type of the DMI determines the helicity and winding direction of the spins, exotic spin textures with various winding numbers and helicities can coexist in frustrated magnets, as seen in Fig. 2.9. In the absence of inversion symmetry breaking, skyrmions (Fig. 2.9b, c) and antiskyrmions (Fig. 2.9a) with opposite winding numbers are degenerate in energy. Additionally, the skyrmions possess a spin degree of freedom (where "spin" refers to the spin of the skyrmion as a particle), associated with a rotation of helicity. In Fig. 2.10a, we show the profile of the mode corresponding to the rotation of helicity in a Néel skyrmion. In the absence of DMI, this mode is a mode of zero energy fluctuations, also known as a Goldstone mode. Goldstone modes and their implications are discussed later in Chap. 3. Bloch-type skyrmions (Fig. 2.9b), Néel-type skyrmions (Fig. 2.9c), and all intermediate helicity states are therefore, in the absence of DMI, degenerate in energy [39]. The skyrmion profile is also different from that of a chiral skyrmion. The polar angle $\theta(\rho)$ goes from $\theta(0) = \pi$ at the centre, to zero, and then oscillates around zero, indicating a reversal of helicity, as shown in Fig. 2.8b. As a consequence, the pairwise interaction potential of skyrmions is non-monotonic as a function of distance, and is found in turn positive (repulsive) and negative (binding) [36, 39], and depends on the relative helicities. It is therefore possible to stabilize a bound (anti)skyrmion/(anti)skyrmion pair (Fig. 2.9f, h, g).

We now focus on the frustrated $J_1 - J_2 - J_3$ magnet on the square lattice under a perpendicular applied field, and $(d, k) = (0, 0)$. The upcoming derivation is based

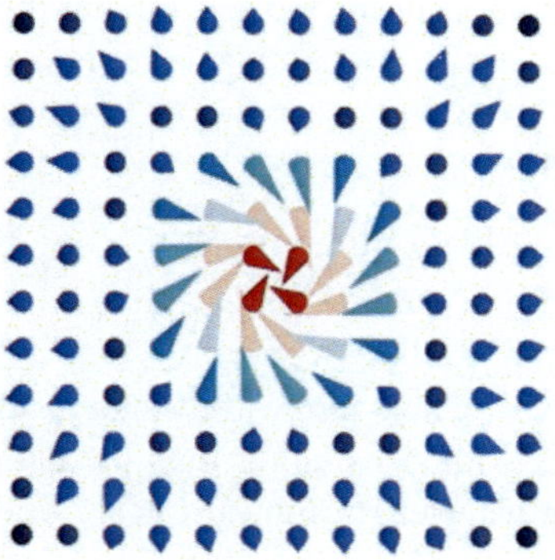
(a)

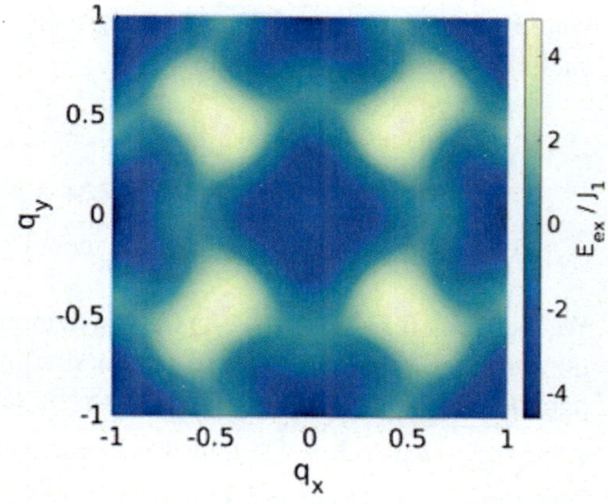

(b)

Fig. 2.10 **a** Profile of the Goldstone mode associated with the rotation of helicity in the Néel skyrmion in the absence of inversion symmetry breaking ($d = 0$). At zero-DMI, Bloch and Néel skyrmions, as well as all skyrmion solutions with intermediate values of the helicity are degenerate in energy. **b** Isotropic exchange energy, $E_{\text{ex}}(\mathbf{q})$, in units of J_1 in the two-dimensional Brillouin zone of the simple square lattice for $(j_2, j_3) = (-0.35, -0.15)$. This particular choice of exchange coupling parameters gives rise to a minimum in the exchange energy around $q_{0x} \approx q_{0y} \approx 0.23$, which corresponds to a SS ground state

on that of Ref. [39]. The competing interactions can stabilize a state with non-zero ordering wave vector $\mathbf{q}$ as a ground state. At zero-temperature and below a critical applied field B_c, the ground state is a homogeneous spin spiral, and the orientation of spin $\mathbf{m}_i$ at site $\mathbf{r}_i$ is expressed as

$$\mathbf{m}_i = \left(\sin\theta\cos(\mathbf{r}_i\cdot\mathbf{q}), \sin\theta\sin(\mathbf{r}_i\cdot\mathbf{q}), \cos\theta\right), \tag{2.42}$$

where the canting angle θ is determined by $\cos\theta = B_z/B_c$. We express the wave vector of the two-dimensional Brillouin zone as

$$\mathbf{q} = \frac{2\pi}{a}\begin{pmatrix} q_x \\ q_y \end{pmatrix}, \tag{2.43}$$

where a is the lattice constant. Setting $\mathbf{m}_i = \mathbf{e}_x$ at position $\mathbf{r}_i = (0, 0)$, its first nearest neighbours are located at

$$\{\mathbf{r}_j\} = \{(a, 0), (-a, 0), (0, a), (0, -a)\}, \tag{2.44}$$

so that the total exchange energy between $\mathbf{m}_i$ and its first neighbours is

$$E_1 = -2J_1\left(\cos(2\pi q_x) + \cos(2\pi q_y)\right), \tag{2.45}$$

where we have chosen $\theta = \pi/2$. Proceeding analogously for the second and third shells of neighbours, we obtain

$$\begin{aligned} E_{\text{ex}}(\mathbf{q}) = &-2J_1\big(\cos(2\pi q_x) + \cos(2\pi q_y)\big) \\ &-2J_2\big(\cos(2\sqrt{2}\pi(q_x + q_y) + \cos(2\sqrt{2}\pi(q_x - q_y)\big) \\ &-2J_3\big(\cos(4\pi q_x) + \cos(4\pi q_y)\big). \end{aligned} \tag{2.46}$$

We then look for a set of (j_2, j_3) for which there exists a non-zero wave vector $\mathbf{q}_0$ that minimizes $E_{\text{ex}}(\mathbf{q_0})$. In order to obtain skyrmions with sizes larger than the lattice parameter, we must limit the search to the large wavelength limit $qa \ll 1$, where $q = \sqrt{q_x^2 + q_y^2}$. In Ref. [39], the exchange energy is expanded in small values of q. The spin lattice introduces a spatial anisotropy, which is minimized by tuning (j_2, j_3). This yields the following reduced coupling parameters: $j_2 = -0.35$; $j_3 = -0.15$, with a minimum in the spin spiral energy around $q_{0x} \approx q_{0y} \approx 0.23$. Figure 2.10b shows the value of $E_{\text{ex}}(\mathbf{q})$ as given in Eq. (2.46) in the first Brillouin zone. The darker area around the center corresponds to the SS ground state. The value of $\mathbf{q}_0$ gives the period of the spin spirals and determines the equilibrium sizes of isolated skyrmion solutions.

2.3.5 Skyrmion Bubbles in Dipolar Magnets

Bubble domains, or skyrmion bubbles [14], originate from the tension of domain walls and surface depolarization as an effect of the sample's shape. In most cases, the profile consists in a spatially extended core with $\theta = \pi$, separated by a domain wall from the uniformly magnetized background where $\theta = 0$. Contrary to chiral skyrmions and skyrmions stabilized by exchange frustration, the core does not exhibit strong localization, with a typically flat or quasi-flat $\theta(\rho)$ profile around the centre [68]. Although bubbles carry an integer topological charge, their typical sizes are found in the micrometer range and, unlike skyrmions, they usually cannot be reduced to nanoscale sizes. We show the spin maps of some bubble skyrmions in Fig. 2.11. In Fig. 2.8c, we show the profile of the Néel bubble in Fig. 2.11b along a radial cross-section. It shows that the core is much more spatially extended than that of the two skyrmions in Fig. 2.8a and b, with a smaller slope in the centre. Although the profile in the center is not flat in this case, this seems to be due to the fact that the bubble is constrained by the 30×30 simulated area, which explains the square shape of the bubble in Fig. 2.11b. The phase diagram in dipolar magnets is similar to that of chiral magnets, with the helical ground state at zero applied field, the bubble lattice at intermediate field and the uniformly magnetized state at higher field, from which metastable isolated bubbles may be stabilized [14, 73]. Contrary to the DMI, the dipolar interactions do not select a single value of the helicity, so that solutions with

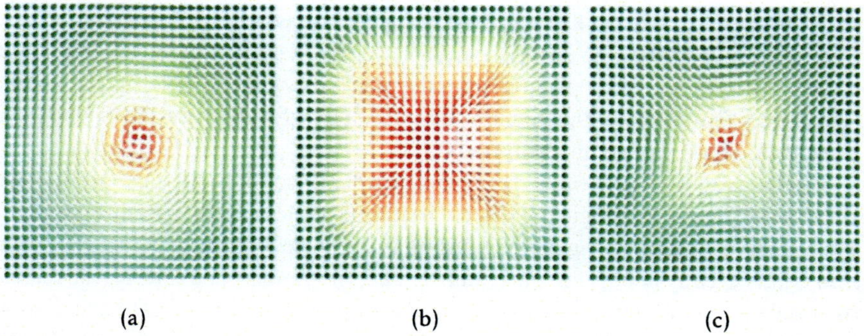

Fig. 2.11 Examples of skyrmion bubbles: **a** Bloch-type bubble with $(n, \phi_0) = (1, 3\pi/2)$, **b** Néel-type bubble with $(n, \phi_0) = (1, \pi)$, **c** antiskyrmion bubble with $(n, \phi_0) = (-1, \pi/2)$

$\phi_0 = \pi/2$ or $3\pi/2$ are degenerate (Fig. 2.11a), and so are solutions with $\phi_0 = 0$ or π (Fig. 2.11b) [35]. Winding numbers of $n = \pm 1$ can also coexist (Fig. 2.11c).

2.4 Summary

In this chapter, we have seen that the main energy contributions in magnets are short-range exchange interactions and long-range dipolar interactions between magnetic moments, as well as interactions of the spins with the crystal lattice and external magnetic fields. The main contributions to the exchange interactions considered in this thesis are the isotropic Heisenberg exchange favouring collinear orientations of neighbouring spins, and the antisymmetric, anisotropic Dzyaloshinskii-Moriya interaction, which favours orthogonal orientations of neighbouring spins with a particular sense of rotation. The DMI arises from spin-orbit coupling in systems with broken inversion symmetry and is responsible for the emergence of noncollinear chiral magnetic orderings. DMI is found in the bulk of noncentrosymmetric crystals such as MnSi, but can also be generated in small centrosymmetric magnetic structures such as thin films, by structural breaking of the inversion symmetry.

On our way to understanding magnetic skyrmions, we have looked at their one-dimensional counterpart in the form of ferromagnetic walls in a chain of spins. Domain walls are a form of topological solitons which arise as solutions to the static sine-Gordon equation in one dimension. In ferromagnets, they extend to higher dimensions between uniformly magnetized domains. Using a variational method, we derived the characteristic quantities in a domain wall, i.e., profile, width, and energy. We made the distinction between Bloch walls, in which the magnetization rotates in the plane of the wall, and Néel walls, in which the magnetization rotates perpendicular to the wall plane. We showed than a chiral pair of π-walls is topologically nontrivial with respect to the easy-plane, that is, if the spins in both walls rotate in the same direction, resulting in a soliton-soliton pair. Nevertheless, within a discretized description of space in which the spins are localized at atomic sites, this configuration can be destroyed by strong enough external effects. Therefore, in reality, topological protection does not hold. Additionally, the strict easy-plane model is not realistic because anisotropies are finite, and so the chiral pair of walls may also be destroyed if the spins rotate in the third dimension.

Lastly, we have presented some characteristics and properties of magnetic skyrmions. These topological solitons arise in a system from the introduction of a lengthscale via competing interactions. In two dimensions of space, they are localized, topologically nontrivial spin textures. If a spin spiral ground state can be realized from only interspin interactions, additional couplings such as magnetic anisotropies and external magnetic fields can then suppress the SS state to recover the uniform FM state, from which isolated skyrmions may be stabilized as metastable excitations. Under an external magnetic field, a skyrmion lattice arises at intermediate field values. Depending on the helicity of the spin texture, we distinguish between Néel and Block skyrmions. Néel skyrmions possess a magnetization profile similar to that of a chiral pair of Néel walls along any radial direction, resulting in a helicity of $\phi_0 = 0$ or π, while Bloch skyrmions possess a profile similar to a chiral pair of Bloch walls, with helicity $\phi_0 = \pi/2$ or $3\pi/2$. Amongst topologically nontrivial spin textures in two dimensions, solutions formally called skyrmions corresponds to the case where the winding direction of the spins matches the spatial travelling direction when travelling around a circular contour at a fixed distance to the skyrmion core, yielding a positive vorticity, $n > 0$. In the case in which the spins rotate in the direction opposite to the travelling direction, the spin texture is an antiskyrmion with negative vorticity, $n < 0$. In that case, the magnetization profile along a radial direction alternates between Néel-type and Bloch-type.

Chiral magnetic skyrmions are skyrmions stabilized by the DMI, and are therefore characterized by a given sense of rotation of the spins along a radial direction. The type of the DMI also selects a particular category of skyrmion solutions. Néel skyrmions are stabilized by interfacial DMI, which is typically found in systems such as Ir/Co/Pt, Pt/Co/AlO_x, and Pd/Fe/Ir(111). Bloch skyrmions are stabilized by bulk DMI, reported, for instance, in the bulk of MnSi, and antiskyrmions are stabilized by anisotropic DMI, which was reported in tetragonal Heusler compounds. Chiral skyrmions possess a strongly localized core, with linear variations of the polar angle along the radial direction, while the asymptotic profile of the tail far from the centre resembles that of a ferromagnetic domain wall.

Another stabilization mechanism of skyrmions is found in frustrated Heisenberg exchange, in which case, for instance, the exchange is ferromagnetic between first neighbours, but antiferromagnetic beyond first neighbours. Skyrmions stabilized by frustrated exchange possess a helicity degree of freedom, and in the absence of DMI, skyrmions and antiskyrmions are degenerate in energy. The skyrmion core is strongly localized, while, in the tail, the polar angle oscillates as a function of distance, indicating a reversal of helicity. While the interaction potential between chiral skyrmions is only repulsive, the oscillations of the polar angle entails that the pairwise interaction potential between skyrmions in frustrated magnets alternates between repulsive and attractive.

Finally, skyrmion bubbles, or bubbles domains, can be stabilized by dipolar couplings. Although they carry a topological charge, they do not exhibit the same topological properties as skyrmions and tend to be much bigger, with a weakly localized core.

In the rest of this thesis, we will focus on ultrathin, perpendicularly magnetized films, so that, in simulations, we can restrict ourselves to two spatial dimensions, and the dipolar energy contribution may be taken as an effective anisotropy. In thin magnetic films (Co, Fe), the breaking of the structural inversion symmetry at an interface with a heavy-metal exhibiting a strong SOC (Ir, Pd) gives rise to an interfacial type of DMI, which in turns favours cycloidal magnetic textures and Néel skyrmions. This is the type of system currently most envisioned for future spintronics applications, in which isolated magnetic skyrmions are to be used as movable data bits. For such technological applications, a precise knowledge and understanding of the average lifetime of isolated skyrmions at room temperature is necessary. Hence, in what follows, we tackle the task of estimating the thermal stability of metastable magnetic skyrmions, with particular emphasis on nanoscale chiral skyrmions stabilized, in most cases, by interfacial DMI. The material parameters used in Chaps. 4 and 5 of the present thesis partially correspond to Pt/Co/AlO_x samples, which is one of the extensively studied systems where nanoscale skyrmions have been experimentally observed. Additionally, frustrated exchange was predicted to arise at certain types of interfaces by DFT calculations, most notably in Pd_{fcc}/Fe, Rh/Fe, and Rh/Co bilayers on Ir(111). Therefore, we shall also study the thermal stability of skyrmions in a frustrated magnet in Chap. 6, as it may well be relevant for applications.

References

1. Craik D (1995) Magnetism: principles and applications. Wiley
2. Landau L, Lifshitz E (1935) Domain walls. Physikalische Zeitschrift der Sowjetunion 8:153
3. Heisenberg W (1928) Zür theorie des ferromagnetismus. Zeitschrift für Physik A Hadrons and Nuclei 49:619–636
4. Dirac PAM (1929) Quantum mechanics of many-electron systems. In: Proceedings of the royal society of London. Series A, Containing papers of a mathematical and physical character, vol 123(792), pp 714–733
5. Dzyaloshinskii IE (1958) A thermodynamic theory of "weak" ferromagnetism of antiferromagnetics. J Phys Chem Solids 4(4):241–255
6. Moriya T (1960) Anisotropic superexchange interaction and weak ferromagnetism. Phys Rev 120(1):91–98
7. Ishikawa Y, Tajima K, Bloch D, Roth M (1976) Helical spin structure in manganese silicide MnSi. Solid State Commun 19(6):525–528
8. Ishikawa Y, Shirane G, Tarvin J, Kohgi M (1977) Magnetic excitations in the weak itinerant ferromagnet MnSi. Phys Rev B 16(11):4956
9. Fert A (1990) Magnetic and transport properties of metallic multilayers. In: Materials science forum, vol 59, pp 439–480. Trans Tech Publications
10. Crépieux A, Lacroix C (1998) Dzyaloshinsky-Moriya interactions induced by symmetry breaking at a surface. J Magn Magn Mater 182(3):341–349
11. Bogdanov AN, Rößler UK (2001) Chiral symmetry breaking in magnetic thin films and multilayers. Phys Rev Lett 87:037203

12. Heinze S, Von Bergmann K, Menzel M, Brede J, Kubetzka A, Wiesendanger R, Bihlmayer G, Blügel S (2011) Spontaneous atomic-scale magnetic skyrmion lattice in two dimensions. Nat Phys 7(9):713
13. Kittel C (1996) Introduction to solid state physics, 7th edn. Wiley
14. Hubert A, Schäfer R (2008) Magnetic domains: the analysis of magnetic microstructures. Springer, Berlin, Heidelberg, Germany
15. Winter JM (1961) Bloch wall excitation. Application to nuclear resonance in a bloch wall. Phys Rev 124:452–459
16. Bloch F (1932) Zur Theorie des Austauschproblems und der Remanenzerscheinung der Ferromagnetika [On the theory of the exchange problem and the remanence phenomenon of ferromagnets]. Zeitschrift für Physik 74:295
17. Tarasenko S, Stankiewicz A, Tarasenko V, Ferré J (1998) Bloch wall dynamics in ultrathin ferromagnetic films. J Magn Magn Mater 189(1):19–24
18. Heide M, Bihlmayer G, Blügel S (2008) Dzyaloshinskii-Moriya interaction accounting for the orientation of magnetic domains in ultrathin films: Fe/W (110). Phys Rev B 78(14):140403
19. Thiaville A, Rohart S, Jué É, Cros V, Fert A (2012) Dynamics of Dzyaloshinskii domain walls in ultrathin magnetic films. Europhys Lett 100(5):57002
20. Zang J, Cros V, Hoffmann A (2018) Topology in magnetism. Springer series in solid-state sciences, vol 192. Springer Nature, Switzerland
21. Braun HB (1994) Fluctuations and instabilities of ferromagnetic domain-wall pairs in an external magnetic field. Phys Rev B 50(22):16485
22. Skyrme THR (1961) A non-linear field theory. Proc R Soc 260:127
23. Piette BM, Schroers BJ, Zakrzewski W (1995) Multisolitons in a two-dimensional Skyrme model. Zeitschrift für Physik C Particles and Fields 65(1):165–174
24. Weidig T (1999) The baby Skyrme models and their multi-skyrmions. Nonlinearity 12(6):1489
25. Schwarz AS (2013) Topology for physicists. A series of comprehensive studies in mathematics, vol 308. Springer, Berlin, Heidelberg, Germany
26. Belavin A, Polyakov A (1975) Metastable states of two-dimensional isotropic ferromagnets. JETP Lett 22(10):245–248
27. Bogdanov A, Yablonskii D (1989) Thermodynamically stable "vortices" in magnetically ordered crystals. The mixed state of magnets. Zhurnal Éksperimental'noĭ i Teoreticheskoĭ Fiziki 95:182
28. Ivanov B, Stephanovich V, Zhmudskii A (1990) Magnetic vortices - the microscopic analogs of magnetic bubbles. J Magn Magn Mater 88(1–2):116–120
29. Bogdanov A, Hubert A (1994) Thermodynamically stable magnetic vortex states in magnetic crystals. J Magn Magn Mater 138(3):255–269
30. Chikazumi S, Graham CD (2009) Physics of ferromagnetism. International series of monographs on physics, vol 94, 2nd edn. Oxford University Press
31. Malozemoff A, Slonczewski J (1979) Magnetic domain walls in bubble materials: advances in materials and device research, vol 1. Academic Press
32. Yu X, Mostovoy M, Tokunaga Y, Zhang W, Kimoto K, Matsui Y, Kaneko Y, Nagaosa N, Tokura Y (2012) Magnetic stripes and skyrmions with helicity reversals. Proc Natl Acad Sci 109(23):8856–8860
33. Finazzi M, Savoini M, Khorsand A, Tsukamoto A, Itoh A, Duo L, Kirilyuk A, Rasing T, Ezawa M (2013) Laser-induced magnetic nanostructures with tunable topological properties. Phys Rev Lett 110(17):177205
34. Nagao M, So Y-G, Yoshida H, Isobe M, Hara T, Ishizuka K, Kimoto K (2013) Direct observation and dynamics of spontaneous skyrmion-like magnetic domains in a ferromagnet. Nat Nanotechnol 8(5):325
35. Koshibae W, Nagaosa N (2016) Theory of antiskyrmions in magnets. Nat Commun 7:10542
36. Leonov A, Mostovoy M (2015) Multiply periodic states and isolated skyrmions in an anisotropic frustrated magnet. Nat Commun 6:8275
37. Okubo T, Chung S, Kawamura H (2012) Multiple-q states and the skyrmion lattice of the triangular-lattice Heisenberg antiferromagnet under magnetic fields. Phys Rev Lett 108(1):017206

38. Nagaosa N, Tokura Y (2013) Topological properties and dynamics of magnetic skyrmions. Nat Nanotechnol 8(12):899–911
39. Lin S-Z, Hayami S (2016) Ginzburg-Landau theory for skyrmions in inversion-symmetric magnets with competing interactions. Phys Rev B 93(6):064430
40. Rózsa L, Palotás K, Deák A, Simon E, Yanes R, Udvardi L, Szunyogh L, Nowak U (2017) Formation and stability of metastable skyrmionic spin structures with various topologies in an ultrathin film. Phys Rev B 95(9):094423
41. Zhang X, Xia J, Zhou Y, Liu X, Zhang H, Ezawa M (2017) Skyrmion dynamics in a frustrated ferromagnetic film and current-induced helicity locking-unlocking transition. Nat Commun 8(1):1717
42. Heo C, Kiselev NS, Nandy AK, Blügel S, Rasing T (2016) Switching of chiral magnetic skyrmions by picosecond magnetic field pulses via transient topological states. Sci Rep 6:27146
43. Berg B, Lüscher M (1981) Definition and statistical distributions of a topological number in the lattice O (3) σ-model. Nucl Phys B 190(2):412–424
44. Böttcher M, Heinze S, Egorov S, Sinova J, Dupé B (2018) B-T phase diagram of Pd/Fe/Ir (111) computed with parallel tempering Monte Carlo. New J Phys 20(10):103014
45. Rohart S, Miltat J, Thiaville A (2016) Path to collapse for an isolated Néel skyrmion. Phys Rev B 93(21):214412
46. Mühlbauer S, Binz B, Jonietz F, Pfleiderer C, Rosch A, Neubauer A, Georgii R, Böni P (2009) Skyrmion lattice in a chiral magnet. Science 323(5916):915–919
47. Yu X, Onose Y, Kanazawa N, Park J, Han J, Matsui Y, Nagaosa N, Tokura Y (2010) Real-space observation of a two-dimensional skyrmion crystal. Nature 465(7300):901
48. Fert A, Cros V, Sampaio J (2013) Skyrmions on the track. Nat Nanotechnol 8(3):152
49. Iwasaki J, Mochizuki M, Nagaosa N (2013) Current-induced skyrmion dynamics in constricted geometries. Nat Nanotechnol 8(10):742
50. Zhou Y, Ezawa M (2014) A reversible conversion between a skyrmion and a domain-wall pair in a junction geometry. Nat Commun 5:4652
51. Müller J (2017) Magnetic skyrmions on a two-lane racetrack. New J Phys 19(2):025002
52. Hsu P-J, Kubetzka A, Finco A, Romming N, von Bergmann K, Wiesendanger R (2017) Electric-field-driven switching of individual magnetic skyrmions. Nat Nanotechnol 12(2):123
53. Romming N, Hanneken C, Menzel M, Bickel JE, Wolter B, von Bergmann K, Kubetzka A, Wiesendanger R (2013) Writing and deleting single magnetic skyrmions. Science 341(6146):636–639
54. Tokunaga Y, Yu X, White J, Rønnow HM, Morikawa D, Taguchi Y, Tokura Y (2015) A new class of chiral materials hosting magnetic skyrmions beyond room temperature. Nat Commun 6:7638
55. Nayak AK, Kumar V, Ma T, Werner P, Pippel E, Sahoo R, Damay F, Rößler UK, Felser C, Parkin SS (2017) Magnetic antiskyrmions above room temperature in tetragonal Heusler materials. Nature 548(7669):561
56. Kanazawa N, Seki S, Tokura Y (2017) Noncentrosymmetric magnets hosting magnetic skyrmions. Adv Mater 29(25):1603227
57. Moreau-Luchaire C, Moutafis C, Reyren N, Sampaio J, Vaz C, Van Horne N, Bouzehouane K, Garcia K, Deranlot C, Warnicke P et al (2016) Additive interfacial chiral interaction in multilayers for stabilization of small individual skyrmions at room temperature. Nat Nanotechnol 11(5):444–448
58. Miron IM, Moore T, Szambolics H, Buda-Prejbeanu LD, Auffret S, Rodmacq B, Pizzini S, Vogel J, Bonfim M, Schuhl A et al (2011) Fast current-induced domain-wall motion controlled by the Rashba effect. Nat Mater 10(6):419
59. Rohart S, Thiaville A (2013) Skyrmion confinement in ultrathin film nanostructures in the presence of Dzyaloshinskii-Moriya interaction. Phys Rev B 88(18):184422
60. Romming N, Kubetzka A, Hanneken C, von Bergmann K, Wiesendanger R (2015) Field-dependent size and shape of single magnetic skyrmions. Phys Rev Lett 114(17):177203
61. Dupé B, Hoffmann M, Paillard C, Heinze S (2014) Tailoring magnetic skyrmions in ultra-thin transition metal films. Nat Commun 5:4030

62. Bessarab PF, Müller GP, Lobanov IS, Rybakov FN, Kiselev NS, Jónsson H, Uzdin VM, Blügel S, Bergqvist L, Delin A (2018) Lifetime of racetrack skyrmions. Sci Rep 8(1):1–10
63. von Malottki S, Bessarab PF, Haldar S, Delin A, Heinze S (2019) Skyrmion lifetime in ultrathin films. Phys Rev B 99(6):060409
64. Banerjee S, Rowland J, Erten O, Randeria M (2014) Enhanced stability of skyrmions in two-dimensional chiral magnets with Rashba spin-orbit coupling. Phys Rev X 4(3):031045
65. Hoffmann M, Zimmermann B, Müller GP, Schürhoff D, Kiselev NS, Melcher C, Blügel S (2017) Antiskyrmions stabilized at interfaces by anisotropic Dzyaloshinskii-Moriya interactions. Nat Commun 8(1):308
66. Camosi L, Rohart S, Fruchart O, Pizzini S, Belmeguenai M, Roussigné Y, Stashkevich A, Cherif SM, Ranno L, De Santis M, Vogel J (2017) Anisotropic dzyaloshinskii-moriya interaction in ultrathin epitaxial au/co/w (110). Phys Rev B 95(21):214422
67. Rybakov FN, Kiselev NS (2019) Chiral magnetic skyrmions with arbitrary topological charge. Phys Rev B 99(6):064437
68. Kiselev N, Bogdanov A, Schäfer R, Rößler U (2011) Chiral skyrmions in thin magnetic films: new objects for magnetic storage technologies? J Phys D: Appl Phys 44(39):392001
69. Butenko G (2012) Phenomenological theory of chiral states in magnets with Dzyaloshinskii-Moriya interactions. PhD thesis, Fakultät Mathematik und Naturwissenschaften der Technischen Universität Dresden
70. Kubetzka A, Pietzsch O, Bode M, Wiesendanger R (2003) Spin-polarized scanning tunneling microscopy study of 360 walls in an external magnetic field. Phys Rev B 67(2):020401
71. Romming N, Pralow H, Kubetzka A, Hoffmann M, von Malottki S, Meyer S, Dupé B, Wiesendanger R, von Bergmann K, Heinze S (2018) Competition of Dzyaloshinskii-Moriya and higher-order exchange interactions in Rh/Fe atomic bilayers on Ir(111). Phys Rev Lett 120:207201
72. Meyer S, Perini M, von Malottki S, Kubetzka A, Wiesendanger R, von Bergmann K, Heinze S (2019) Isolated zero field sub-10 nm skyrmions in ultrathin Co films. Nat Commun 10:3823
73. Garel T, Doniach S (1982) Phase transitions with spontaneous modulation-the dipolar Ising ferromagnet. Phys Rev B 26(1):325

Chapter 3
Langer's Theory and Application to Magnetic Spin Systems

3.1 Introduction

3.1.1 Preamble: On the Theory of Brownian Motion

The first account of Brownian motion is attributed to the botanist Robert Brown in 1827 [1] who observed aqueous suspensions of pollen grains under a microscope, and found that the grains were performing "*rapid oscillatory motion*". Throughout the nineteenth century, this motion of a so-called Brownian particle immersed in a solvent sparked many investigations and speculations regarding its origin, and the following properties emerged:

- the trajectories are highly irregular and non-differentiable;
- the motion is independent of the nature of the particle: it is a universal phenomenon of mechanical origin;
- the motion never ceases as constant energy is injected into it, and thus pertains to the realm of out-of-equilibrium dynamics;
- the smaller the particles, the higher the temperature, and the higher the viscosity of the fluid, the more active the motion.

As is well-known, it was Einstein who, in 1905, initiated the theory of Brownian motion by combining the random walk process with the Maxwell-Boltzmann distribution [2]. He attributed the cause of the motion to the thermal fluctuations of the molecules of the solvent. In a frictionless fluid, a collision with a molecule of the solvent would modify the velocity of the particle. However, in a viscous fluid, the change in velocity is rapidly dissipated, and the result is a change in the position of the particle. Einstein assumed that the overall result of the collisions is a kind of random walk of the particle. The interplay of the injection of energy through collision, followed by its dissipation through the fluid viscosity, is an example of what is known as the fluctuation-dissipation theorem [3].

L. Desplat, *Thermal Stability of Metastable Magnetic Skyrmions*, Springer Theses,
https://doi.org/10.1007/978-3-030-66026-0_3

Consider a small particle of mass M immersed in a solvent at temperature T. The following discussions are carried out in one dimension without loss of generality, and the extension to higher dimensions is straightforward. Formally, the microscopic treatment of the system would involve the Brownian particle and all the molecules of the solvent. This is an N-body problem, where $N \approx 10^{23}$, which therefore cannot be solved analytically. From the equipartition theorem in one dimension, the mean energy of the particle is [4]

$$\frac{1}{2}M\langle v^2\rangle = \frac{1}{2}k_B T, \tag{3.1}$$

where v is the velocity of the particle, k_B is Boltzmann's constant, and T is the absolute temperature. Let the molecules of the solvent be of mass m, with $m \ll M$, then the thermal velocity of the Brownian particle satisfies

$$v_{BP} = \sqrt{\langle v^2\rangle} = \sqrt{k_B T/M} = v_{\text{solvent}}\sqrt{m/M} \ll v_{\text{solvent}}. \tag{3.2}$$

If follows that the position, x, of the particle is a slow variable on the timescale of the solvent. Einstein introduced a "coarse-graining" which eliminates the rapid degrees of freedom of the solvent and leads to the apparent stochastic nature of the motion. The solvent is then treated as a structureless medium with only a temperature T. Stochasticity emerges because, if we were to consider a second system of the particle immersed in the fluid, identical to the first system in all aspects but the initial conditions of the molecules in the fluid, then the motion of the particle would be different. We consider an ensemble of such systems, known as a Gibbs ensemble. While we cannot predict the time-evolution of the (stochastic) macroscopic observables of the system, we can predict that of ensemble averages. For instance, the probability to find the position of the particle in the interval $(x, x+dx)$ is equivalent to the number of systems of the ensemble with positions found in the interval $(x, x+dx)$ divided by the number of systems in the ensemble. Since x is a continuous variable, let $W(x,t)$ be a probability density, also known in the literature as a probability distribution function of the variable x. Its time evolution obeys a class of equations known as the Fokker-Planck equation (FPE), which, in one dimension, takes the form [4]

$$\frac{\partial W}{\partial t} = \left[\frac{\partial}{\partial x}D^{(1)}(x) + \frac{\partial^2}{\partial x^2}D^{(2)}(x)\right]W. \tag{3.3}$$

In Eq. (3.3), $D^{(1)}(x)$ is the drift coefficient, of deterministic origin, and $D^{(2)}(x)$ is the diffusion coefficient, of stochastic origin. The FPE represents an equation of motion for the distribution function.

In 1908, Langevin [5] postulated that the velocity $v = dx/dt = \dot{x}$ must also be a slow variable on the timescale of the solvent, and thus proposed a mechanistic description of Brownian motion. A Brownian particle in a solvent will, first of all, feel a friction force, whose expression is given by Stokes' law, $F_c = -\zeta v$, where ζ is the friction coefficient. Furthermore, Langevin introduced a stochastic fluctuating force $F_f(t)$ in the form of Gaussian white noise. That force contains all contributions from the rapid degrees of freedom eliminated by the coarse-graining. Using Newton's second law of motion and dividing by the mass M of the particle, we arrive at Langevin's equation for the motion of a free Brownian particle,

$$\dot{v} = -\gamma v + \Gamma(t), \tag{3.4}$$

where $\dot{v} = dv/dt$, $\gamma = \zeta/M$, and $\Gamma(t) = F_f(t)/M$ is a fluctuating force per unit mass called the Langevin force. Its statistical properties are as follows. Firstly,

$$\langle \Gamma(t) \rangle = 0, \tag{3.5}$$

because on average, $\langle v(t) \rangle$ should obey the deterministic version of Eq. (3.4) obtained by setting $\Gamma(t) = 0$. The symbol $\langle \ldots \rangle$ corresponds to an ensemble average. Secondly, we suppose that the collisions with different molecules of the fluid should be independent, so that if we multiply two Langevin forces at different times, t and t', the average value is zero if the time difference $| t - t' |$ is larger than the average duration of a collision, τ_0, i.e., $\tau_0 \ll | t - t' |$. Furthermore, τ_0 is usually much smaller than the relaxation time $\tau = 1/\gamma$, and so we take the limit $\tau_0 \to 0$, which yields,

$$\langle \Gamma(t)\Gamma(t') \rangle = 2\gamma k_B T \delta(t - t'), \tag{3.6}$$

where $\delta(t - t')$ is the Dirac distribution. Since $\Gamma(t)$ is a Gaussian process, it is completely characterized by (3.5) and (3.6). We can note that, although Einstein assumed that the trajectory is non-differentiable, Langevin could define a velocity v by considering a finer timescale for the coarse-graining. If the noise $\Gamma(t)$ does not depend on x, it is called additive. On the other hand, if it does depend on x, it is called multiplicative noise, and requires an appropriate interpretation.

3.1.2 Reaction Rate Theory

Modern reaction rate theory initially developed independently from the theory of Brownian motion. Its origin dates back to the 1880s, when Arrhenius [6] first described the rate coefficient of a chemical reaction as

$$k = f_0 e^{-\Delta E/k_B T}, \tag{3.7}$$

where ΔE is the activation energy, and f_0 is a prefactor. This equation subsequently led to the representation of chemical reactions as an ensemble of particles situated at the bottom of a potential well A (Fig. 3.1a). Over time, under the influence of thermal agitation, rare particles may escape over a barrier of height $\Delta E \gg k_B T$ at S, and never return to A. In the context of transition state theory (TST), the Arrhenius law is expressed as [6],

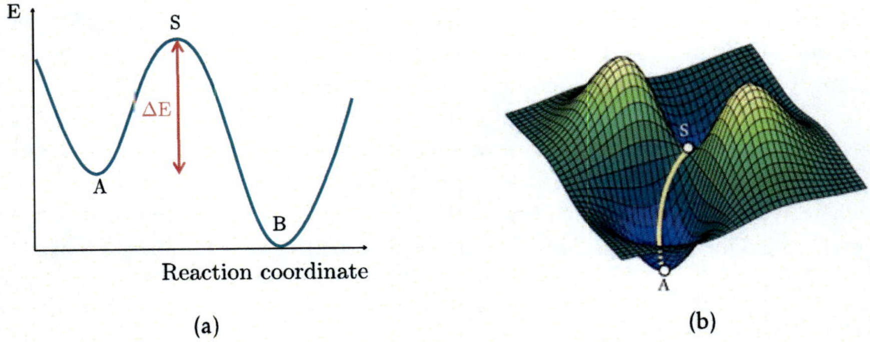

Fig. 3.1 Example energy surface of a system possessing a metastable local minimum A and a lower energy minimum B, separated by an energy barrier at S. The case **a** can be viewed as the one dimensional system treated by Kramers, or a projection of **b** along the reaction coordinate. The case **b** is an example energy landscape for a system possessing two degrees of freedom. In that case, the top of the barrier is a saddle point. The reaction coordinate corresponds to the path of minimum energy connecting A, S and M. If the saddle point is of first order, it constitutes a maximum of the energy along the reaction coordinate, but a minimum along all the other degrees of freedom

$$k_{\mathrm{TST}} = \frac{\omega_A}{2\pi} e^{-\beta \Delta E}, \tag{3.8}$$

where $\beta = 1/k_B T$ is Boltzmann's constant. The attempt frequency ω_A is the angular frequency of a particle performing oscillatory motion at the bottom of the potential well. The exponential Boltzmann factor weighs the escape from the well.

Reaction rate theory was set in the context of non-equilibrium statistical mechanics in 1940 by the work of Kramers [7]. He chose to model a chemical reaction as a classical particle moving in a one dimensional potential $E(x)$, in which x is the reaction coordinate. The particle is coupled to its environment in the form of a heat bath which remains in perpetual thermal equilibrium at temperature T. The coupling to the heat bath is modelled by the Brownian motion performed by the particle [2], which represents all the remaining degrees of freedom of the total system consisting of the particle and the heat bath. Kramers assumed that a (Gibbs) ensemble of particles are initially trapped in the potential well near A. They receive energy from their surroundings, such that a Maxwell-Boltzmann distribution is rapidly attained in the well. That is, $W(x, t) \sim e^{-\beta E(x)}$, so that, in other words, the number of particles with energy found between E and $E + dE$ is proportional to $e^{-\beta E(x)} dE$. Over time, rare particles may gain energy in excess of the barrier height and escape over the barrier at S to reach the lower energy minimum B, never to return. The diffusion over the barrier is assumed to be a quasi-stationary process, so that the disturbance to the Maxwell-Boltzmann distribution can be neglected at all times. The rate constant corresponds to the average number of particles of the ensemble that cross the barrier per unit time. Kramers' objective was to derive a prefactor from a microscopic model of the reaction system, and insert it into the TST result. The Kramers formula for the escape from an *isolated* well (no returns from B) takes the form,

$$k_{\mathrm{Kr}} = A \frac{\omega_A}{2\pi} e^{-\beta \Delta E}, \qquad (3.9)$$

where A contains non-equilibrium effects caused by the disturbance to the Maxwell-Boltzmann distribution at the barrier top, and describes deviations from the TST rate in Eq. (3.8). That model is purely classical, and yields explicit formulas for the rate for very low, and intermediate-to-high dissipative coupling to the heat bath.

In 1969, Langer [8] proposed the most complete treatment of the extension of Kramers theory to a multidimensional phase space in the intermediate-to-high damping (IHD) regime [9]. His treatment is a generalization of a 1935 calculation by Becker and Döring of the rate of condensation of a supersaturated vapour [10]. According to Langer, if the characteristic disturbance that leads to the transition appears spontaneously as a thermodynamic fluctuation, it is an intrinsic property of the system and is classified as a homogeneous disturbance. On the other hand, if it is caused by a foreign object that is not part of the system, such as the presence of defects in the wall of the container, it is classified as inhomogeneous, of lesser fundamental interest. It is the former that was treated by Langer in his statistical theory for the decay of metastable states. The extension to many degrees of freedom makes the theory applicable to magnetic spin systems, with interactions such as exchange and dipole-dipole couplings. Within a multidimensional energy surface, (meta)stable states correspond to (local) energy minima with respect to all degrees of freedom. The internal energy barrier, ΔE, is linked to the most probable path(s) through the energy landscape, in the form of minimum energy paths (MEPs), which usually traverse a first-order saddle point (SP). First-order saddle points represent a minimum of the energy with respect to all degrees of freedom, except one—the reaction coordinate, which connects states A, S, and B. This is illustrated in Fig. 3.1b. If several saddle points are present in the energy landscape, and the different transition processes can be considered independent from one another, the total escape rate out of A is a priori given by the sum of the escape rates over each saddle point.

The exponential Boltzmann factor in Eq. (3.7) shows that the decay of a metastable state takes place over time scales which are much longer than the time scales linked with the intrinsic dynamics of the system. For this reason, solely understanding the dynamics of a system is not enough in order to predict the process(es) by which it decays, and it is essential to study and understand the decay mechanisms themselves. For systems with many degrees of freedom and many-body interactions such as magnetic spin systems, this is often an arduous task, and relying on numerical schemes becomes almost unavoidable. The difficulty in computing transition rates for such a class of systems thus lies in the precise identification of the first-order saddle point(s) on the energy landscape on the one hand, and on the correct evaluation of the rate prefactor, f_0, on the other hand.

In this chapter, we firstly give the scope of Langer's theory and how it can be applied to magnetic spins systems in Sect. 3.2. In that context, we detail the calculation of the main contributions entering in the rate prefactor. We then describe the geodesic nudged elastic band (GNEB) method [11] in Sect. 3.4, which enables the numerical computation of a minimum energy path on the complex multivariable energy surface, and the precise identification of the first-order saddle point on the path. Finally, in Sect. 3.5, we give details of the numerical implementation of the different building blocks in Langer's theory on the atomistic spin lattice.

3.2 Langer's Theory in the IHD Regime

3.2.1 Framework for the Derivation of the Rate Expression

Langer's theory describes the rate of decay of a metastable state under the effect of thermal fluctuations, provided certain conditions are fulfilled. It is set within the intermediate-to-high damping (IHD) regime, and the other main assumptions essentially mirror that of Kramers [7]. The barrier height must be large compared to thermal fluctuations, $\Delta E \gg k_B T$, so that the diffusion over the barrier is slow enough as to ensure that a Maxwell-Boltzmann distribution is rapidly attained and maintained in the metastable well at all times. It follows that the contribution to the flux over the saddle point will come mainly from a small region around S [9]. This also ensures that barrier re-crossing events are negligible, so that state B is viewed as a sink of probability. According to Ref. [9], $\Delta E \geq 5k_B T$ seems to be a reasonable lower bound for the energy barrier. Additionally, we postulate small amplitudes of the fluctuations about the extrema states A and S, so that the energy surface in their vicinity can be approximated as a quasiharmonic potential. This assumption becomes more accurate at low T.

The Hamiltonian $E(\boldsymbol{\eta})$ is function of $2N$ variables $\boldsymbol{\eta} = \{\eta_1, \eta_2, \dots \eta_{2N}\}$, which could be coordinates and associated conjugate momenta of a point in phase space, or coordinates describing the orientation of magnetic vectors. $E(\boldsymbol{\eta})$ possesses two minima, A and B, separated by a saddle point S, and $E(\boldsymbol{\eta}^A) > E(\boldsymbol{\eta}^B)$. We assume that the energy in the vicinity of the extrema states can be approximated as a Taylor series truncated to the second order term,

$$E(\boldsymbol{\eta}) \sim E^0(\tilde{\boldsymbol{\eta}}) + \frac{1}{2}(\boldsymbol{\eta} - \tilde{\boldsymbol{\eta}}) H_{\tilde{\boldsymbol{\eta}}} (\boldsymbol{\eta} - \tilde{\boldsymbol{\eta}})^T, \tag{3.10}$$

where $\tilde{\boldsymbol{\eta}} = (\tilde{\eta}_1 \dots \tilde{\eta}_{2N})$ are the coordinates of a local extremum (the local minimum A or saddle point S). At the extrema,

$$\frac{\partial E}{\partial \boldsymbol{\eta}}\Big|_{\tilde{\boldsymbol{\eta}}} = 0, \tag{3.11}$$

in which the notation $|_{\tilde{\boldsymbol{\eta}}}$ means the expression is evaluated at $\tilde{\boldsymbol{\eta}}$, and

$$H_{\tilde{\eta}} = \begin{pmatrix} \frac{\partial^2 E}{\partial \eta_1^2}\Big|_{\tilde{\eta}} & \cdots & \frac{\partial^2 E}{\partial \eta_1 \partial \eta_{2N}}\Big|_{\tilde{\eta}} \\ \vdots & & \vdots \\ \frac{\partial^2 E}{\partial \eta_{2N} \partial \eta_1}\Big|_{\tilde{\eta}} & \cdots & \frac{\partial^2 E}{\partial \eta_{2N}^2}\Big|_{\tilde{\eta}} \end{pmatrix}, \tag{3.12}$$

is the energy Hessian evaluated at $\tilde{\eta}$, which contains the second derivatives of the energy with respect to all degrees of freedom. It is symmetric and real, and therefore Hermitian by construction, which ensures that it admits only real eigenvectors and eigenvalues. The Hessian can be diagonalized by expanding small fluctuations about the extrema in the eigenbasis,

$$\eta_i - \tilde{\eta}_i = a_i \chi_i, \tag{3.13}$$

where $\{a_i\}$, $i = 1 \ldots 2N$ is a set of eigenbasis coordinates characterizing the amplitudes of the fluctuations, and $\{\chi_i\}$ is a set of orthonormal eigenvectors forming a complete basis in the space of configurations. They satisfy the eigenvalue equation,

$$H \chi_i = \lambda_i \chi_i, \tag{3.14}$$

where the $\{\lambda_i\}$ are the eigenvalues of the Hessian. Because H is evaluated at energy extrema, the $\{\lambda_i\}$ correspond to the $2N$ curvatures of the energy landscape. From Eqs. (3.10), (3.13), and (3.14), the energy of small deviations from an extremum can be linearized as

$$E(\boldsymbol{\eta}) \sim E^0(\tilde{\boldsymbol{\eta}}) + \frac{1}{2}\sum_i \lambda_i a_i^2. \tag{3.15}$$

In what follows, we mainly follow the notations from [6, 9, 12]. The following assumptions are made about the density of states of the system, $\rho(\boldsymbol{\eta}, t)$.

1. The time evolution of of the density of states $\rho(\boldsymbol{\eta}, t)$ obeys a Fokker-Plank equation,

$$\frac{\partial \rho(\boldsymbol{\eta}, t)}{\partial t} = \sum_{i=1}^{2N}\sum_{j=1}^{2N} \frac{\partial}{\partial \eta_i} M_{ij}\Big[\frac{\partial E}{\partial \eta_j} + k_B T \frac{\partial}{\partial \eta_j}\Big]\rho(\boldsymbol{\eta}, t), \tag{3.16}$$

where $\mathbf{M} = (M_{ij})$ is the transport matrix, which we assume to be constant. $\mathbf{M}$ has a symmetric part $\mathbf{D} = \frac{1}{2}(\mathbf{M} + \mathbf{M}^T)$, referred to as the diffusion matrix, which characterizes the thermal fluctuations due to the heat bath, and an antisymmetric part $\mathbf{A} = \frac{1}{2}(\mathbf{M} - \mathbf{M}^T)$, which describes the motion in the absence of the heat bath. Equation (3.16) is linearized about the saddle point $\boldsymbol{\eta}^S$ by inserting (3.10) into (3.16), which yields

$$\frac{\partial \rho}{\partial t} = \sum_{ij} \frac{\partial}{\partial \eta_i} M_{ij}\Big[\sum_k H^S_{jk}(\eta_k - \eta^S_k) + k_B T \frac{\partial}{\partial \eta_j}\Big]\rho(\boldsymbol{\eta}), \tag{3.17}$$

in which the H^S_{jk} are entries in the Hessian matrix evaluated at S. The FPE (Eq. 3.17) can be interpreted as a continuity equation of the $2N$-dimensional probability current density $J_i(\boldsymbol{\eta}, t)$ as

$$\frac{\partial \rho}{\partial t} = \sum_i \frac{\partial J_i}{\partial \eta_i}, \tag{3.18}$$

where

$$J_i = \sum_{j=1}^{2N} M_{ij} \Big[\sum_k H^S_{jk}(\eta_k - \eta^S_k) + k_B T \frac{\partial}{\partial \eta_j} \Big] \rho(\boldsymbol{\eta}, t). \tag{3.19}$$

Since we assumed quasi-equilibrium,

$$\frac{\partial \rho}{\partial t} = 0. \tag{3.20}$$

2. The high barrier ensures that, in the vicinity of A, the density of states corresponds to the Maxwell-Boltzmann distribution,

$$\rho(\boldsymbol{\eta}) = \rho_{eq}(\boldsymbol{\eta}) = \frac{1}{Z} e^{-\beta E(\boldsymbol{\eta})}, \qquad \boldsymbol{\eta} \sim \boldsymbol{\eta}^A, \tag{3.21}$$

where

$$Z = \int_{-\infty}^{\infty} \dots \int_{-\infty}^{\infty} e^{-\beta E} d\eta_1 \dots d\eta_{2N} \tag{3.22}$$

is the partition function.

3. Practically no particles have reached the lower minimum S, so we have the sink boundary condition

$$\rho(\boldsymbol{\eta}) = 0, \qquad \boldsymbol{\eta} \text{ beyond } \boldsymbol{\eta}^S. \tag{3.23}$$

Although the system is assumed to be at quasi-equilibrium, there is a weak current of rare particles over the barrier, in which case the density of states is written as

$$\rho(\boldsymbol{\eta}) = \zeta(\boldsymbol{\eta}) \rho_{eq}(\boldsymbol{\eta}). \tag{3.24}$$

ζ is the crossover function which characterizes the diffusion over the barrier. It is expressed as an error function of a single variable u as

$$\zeta(\boldsymbol{\eta}) = \zeta(u) = \frac{1}{2\pi k_B T} \int_u^{\infty} \exp\Big(-\frac{\beta z^2}{2}\Big) dz, \tag{3.25}$$

where u is a linear combination of the η_i's,

$$u = \sum_{i=1}^{2N} U_i(\eta_i - \eta_i^S), \tag{3.26}$$

with the U_i's chosen as constants. The rate k depends on j, the flux of particles over the barrier, and n_A, the number of particles in the metastable well, that is,

$$k = \frac{j}{n_A}. \tag{3.27}$$

The expression for the flux of particles is calculated [8, 9] by integrating the current density (Eqs. (3.18) and (3.17)) through the plane $u = 0$ as

$$j = \sum_i \int_{u=0} J_i(\eta) dS_i. \tag{3.28}$$

u is, in fact, parametrized by a_0, which is the eigenbasis coordinate associated with the unstable mode at the top of the barrier, and corresponds to the reaction coordinate. By convention, the dividing surface defined by $a_0 = 0$ passes through the saddle point, while the metastable (saddle) region is found at negative (positive) values of a_0. This is represented in a schematic view of the energy landscape in Fig. 3.2. The corresponding eigenvector χ_0 gives the direction of the current.

The number of particles in the well, n_A, is calculated by integrating the density of states in the well (Eq. 3.21),

$$\begin{aligned} n_A &= \int \dots \int \rho_{\text{eq}} d\eta_1 \dots d\eta_{2N} \\ &= Z^{-1} \int \dots \int e^{-\beta E_A} d\eta_1 \dots d\eta_{2N}, \end{aligned} \tag{3.29}$$

in which the partition function, Z, was previously defined in Eq. (3.22). In the previous expression, $E_A = E(\eta^A)$ is subsequently linearized according to Eq. (3.15). This term, as well as (3.28), yield products of Gaussian integrals of the form,

$$\prod_{i=1}^{2N} \int_{-\infty}^{+\infty} \exp\left(-\frac{\beta}{2} \lambda_i a_i^2 \right) da_i = \prod_{i=1}^{2N} \sqrt{\frac{2\pi}{\beta \lambda_i}}. \tag{3.30}$$

In the end, the total rate of escape out of state A is expressed as

$$k = \frac{\lambda_+}{2\pi} \Omega_0 e^{-\beta \Delta E} = \frac{\lambda_+}{2\pi} \sqrt{\frac{\det H^A}{|\det H^S|}} e^{-\beta \Delta E}, \tag{3.31}$$

where Ω_0 is the ratio of energy curvatures in the metastable well and at the saddle point, and λ_+ is the dynamical contribution that takes into account the dissipative

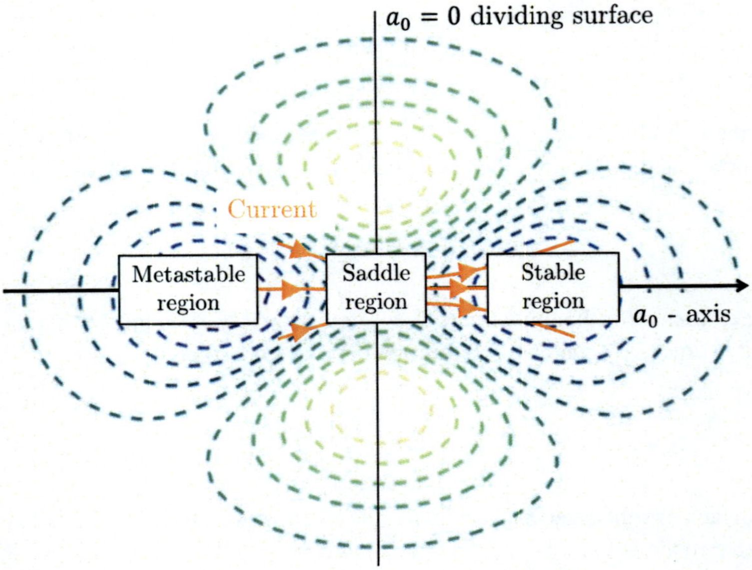

Fig. 3.2 Schematic view of the energy surface where a metastable region located at $a_0 < 0$ is separated from a more stable region ($a_0 > 0$) by a saddle. a_0 is the eigenbasis coordinate associated with the unstable mode at the top of the barrier and represents the reaction coordinate. A weak current of particles moving in the direction of increasing a_0, where each particle represents the entire system, originates from a small region around the saddle point and reaches the more stable region. The rate k of the transition is given by the ratio of the current over the barrier and the number of particles in the metastable region

dynamics of the system at the top of the barrier, and constitutes a correction from TST rates. This results pertains to the IHD regime due to the quasiharmonic approximation of the energy that was performed in the vicinity of the extrema. At high damping, the system relaxes faster and explores less of its phase space. Low damping, on the other hand, implies that the system takes longer to relax back to the equilibrium state after a perturbation, and consequently visits a wider area of its phase space. At very low damping, the region where the density of states deviates from the Maxwell-Boltzmann distribution—i.e. the region where the system deviates from equilibrium—extends beyond the small region around the saddle point in which the potential may be replaced by a quadratic approximation, so the theory fails.

The meaning and derivation of the terms in Eq. (3.31) for magnetic spin systems are discussed in what follows. The escape rate may also be written in terms of the partition functions around the states A and S, respectively Z_A and Z_S as [13]

$$k = \frac{\lambda +}{2\pi} \frac{Z_S}{Z_A}. \tag{3.32}$$

3.2.2 *Ratio of Energy Curvatures: Interpretation and the Case of Goldstone Modes*

When evaluated at an extremum, the $\{\lambda_i\}$ correspond to the $2N$ curvatures of the energy surface in normal mode space. A positive (negative) curvature corresponds to a mode of stable (unstable) fluctuations. A zero-curvature corresponds to a Goldstone mode of zero energy fluctuation, and is associated with a continuous unbroken global symmetry [8, 14]. In the case of a first-order saddle point, all curvatures at A and S are either positive or zero-curvatures, aside from a single negative curvature at the top of the barrier. This unstable mode is the one that will eventually allow the system overcome the barrier and reach the lower energy minimum.

In the absence of Goldstone modes, the factor Ω_0 in Eq. (3.31) is obtained from the square root of the product of the curvatures of the energy at states A and S,

$$\Omega_0 = \sqrt{\frac{\det H^A}{|\det H^S|}} = \sqrt{\frac{\prod_i \lambda_i^A}{\prod_j |\lambda_j^S|}}. \tag{3.33}$$

3.2.2.1 The Case of Zero-Curvatures

If the Hessian contains one or several zero-eigenvalue(s) at A and/or S, the contribution of the zero-mode(s) to the prefactor needs to be handled separately. The zero-eigenvalue is removed from the determinant of the Hessian, and instead of a Gaussian integral (Eq. (3.30)), we get a contribution of the type

$$\int da_l, \tag{3.34}$$

where a_l is the eigenbasis coordinate associated with the zero-eigenvalue λ_l, and the integration is performed over the domain of a_l. If this domain is finite, we get a finite contribution to the prefactor. If we assume we have k zero-eigenvalues in the Hessian, we get,

$$v = \int da_{l_1} \int da_{l_2} \dots \int da_{l_k}, \tag{3.35}$$

where v is the volume of the k-dimensional subspace associated with the variables $\{a_{l_j}\}$, $j = 1 \dots k$, that correspond to zero-eigenvalues. Additionally, a factor $(2\pi/\beta)^{-1/2}$ appears for each Goldstone mode at S, and $(2\pi/\beta)^{1/2}$ appears for each Goldstone mode at A. This is because the $(2\pi/\beta)^{1/2}$ factors in Eq. (3.30) no longer cancel out if the number of Gaussian integrals entering in the final expression of Ω_0 differs between A and S. In other words, if the number of Goldstone modes is non-zero and different at A and S, the attempt frequency obtains a temperature-dependence. Hence, in the more general case where there are k_A zero-curvatures at A and k_S zero-curvatures at S, Eq. (3.33) becomes

$$\Omega_0 = \left(\frac{2\pi}{\beta}\right)^{(k_A-k_S)/2}\sqrt{\frac{\det' H^A}{|\det' H^S|}\frac{v_S}{v_A}}, \tag{3.36}$$

where det′ is a determinant calculated after removing the zero-eigenvalues, and v_A, v_S are the volumes associated with the Goldstone modes at A and S.

Langer's theory treatments of rotational and translational Goldstone modes in magnetic spin systems were previously carried out in works by Braun [14] and Loxley [13]. We can also mention the more recent Refs. [15, 16] where it was done numerically in the context of skyrmion lifetimes. Although the authors used a different version of reaction rate theory, namely Harmonic TST (HTST) [17], the extension of HTST to include Goldstone modes is the same as that of Langer's theory. Equation (3.36) should also come with some form of criterion for when an eigenmode should be treated as a Goldstone mode. Numerically—and experimentally—an energy curvature of exactly zero will in principle not exist. An intuitive idea for such a criterion is that a curvature will appear flat to the system if it is small compared to the amplitude of thermal fluctuations. For translational Goldstone modes, the limit [16, 18]

$$\frac{1}{2}\lambda_{\text{tr}} v^2 \lessapprox k_B T, \tag{3.37}$$

where $\lambda_{\text{tr}} \approx 0$ is an eigenvalue associated with a translation mode, was proposed.

3.2.2.2 Change in Entropy

To interpret the physical meaning behind Ω_0, we examine Eq. (3.33). The product of curvatures evaluated at an extremum in the energy surface, $\left(\prod_i \lambda_i^{A,S}\right)^{-1}$, can be seen as a measure of the total volume of configuration space (η-space) accessible to thermal fluctuations in that particular state. The ratio of eigenvalues thus corresponds to the change in that volume induced by the transition from A to S. In other words, it characterizes the relative volume of the saddle point region. $\Omega_0 \ll 1$ is interpreted as a large volume of the metastable well, and/or a narrow saddle region in η-space (eg: Fig. 3.1b). As a consequence, the probability that the system will visit the saddle region is low. These considerations bring us to the notion of entropy, which measures the number of micro-realisations which exist for a given macrostate, and is also commonly interpreted as a measure of disorder. As entropy is normally defined for a stable equilibrium state, we define the change in configurational entropy ΔS with respect to stable fluctuations only. If the saddle point is of first order, this corresponds to [13]

$$e^{\Delta S/k_B} \equiv \sqrt{\frac{\beta}{2\pi}}\sqrt{\frac{\prod_i \lambda_i^A}{\prod_j' \lambda_j^S}}, \tag{3.38}$$

where $\prod'$ is defined for positive curvatures, and an additional $(\beta/2\pi)^{1/2}$ factor is needed to keep the dimension consistent. The total activation energy corresponds to the change in Helmholtz free energy, and takes into account the change in internal energy, ΔE, and the change in entropy, ΔS, such that $\Delta F = \Delta E - T\Delta S$. It follows that Eq. (3.33) can be expressed as

$$\Omega_0 = \sqrt{\frac{2\pi}{\beta|\lambda_1^S|}} e^{\Delta S/k_B}, \tag{3.39}$$

in which λ_1^S is the negative curvature at S. The factor Ω_0 is thus a measure of the number of available configurations, and gives the entropic contribution to the prefactor. More generally, if the saddle point is of order k, i.e., if k unstable modes exist at S, Eqs. (3.38) and (3.39) may be generalized as

$$e^{\Delta S/k_B} \equiv \left(\frac{\beta}{2\pi}\right)^{k/2} \sqrt{\frac{\prod_i \lambda_i^A}{\prod_j' \lambda_j^S}}, \tag{3.40}$$

and

$$\Omega_0 = \left(\frac{2\pi}{\beta}\right)^{k/2} \sqrt{\prod_j'' |\lambda_j^S|^{-1}} e^{\Delta S/k_B}, \tag{3.41}$$

in which $\prod''$ is a product over negative curvatures.

3.3 Application to Magnetic Spin Systems

In what follows, we consider a system of N magnetic spins on a lattice, for which the change in their magnitude is generally much faster than the change in their orientation. This separation of timescales justifies the assumption that the amplitudes of the moments remain constant. The system can thus be described in terms of orientations of the moments alone, and the energy surface reduces to a $2N$-dimensional landscape. Below, we provide details of the derivation of the different contributions to the rate prefactor of Langer's theory, in the context of its application to a system of magnetic spins. The restriction to the IHD regime means that the scope of the theory is limited to cases where the precessional dynamics can be neglected, in the sense that it does not significantly impact the transition path, and the time scale of the transition is set by the dissipation rate.

One requirement of Kramers' method—and subsequently Langer's theory—is that the system may be modelled by the theory of Brownian motion. For magnetic systems, the environment acts a heat bath of constant temperature T. The magnetic vectors interact with the fast degrees of freedom of the environment in the form of phonons, conducting electrons, nuclear spins, etc. [19], leading to fluctuations in their orientation. The interactions of the spins with the environment may be modelled by a fast, fluctuating, stochastic magnetic field. The equation that governs the dynamics of the magnetic vectors, namely the Landau-Lifshitz-Gilbert equation (Eq. (3.46)), may be cast into a Langevin equation with multiplicative noise by including this term. This will be explored in more details in Chap. 5.

In the past, Langer's theory was successfully applied to the analytical treatment of incoherent magnetization reversals by Braun in nanowires [14], and by Loxley and Stamps in soft/hard nanowires [20]. The first numerical implementations of the theory for magnetic spin systems are more recent and were performed in conjunction with the finite element method [21] for magnetization reversal problems in graded media grains.

3.3.1 Construction of the Energy Hessian in Spherical Coordinates

The total energy of the system may be expressed in terms of spherical coordinates on the unit sphere $E(\boldsymbol{\theta}, \boldsymbol{\phi})$, where $\boldsymbol{\theta} = (\theta_1 \ldots \theta_N)$ is the polar angle with respect to the Cartesian z-axis and $\boldsymbol{\phi} = (\phi_1 \ldots \phi_N)$ is the corresponding azimuth in the xy-plane (see Sect. 2.1). In principle, it is necessary to define canonically conjugate variables $(\mathbf{p}, \mathbf{q})$ [8, 9, 21], such that

$$\begin{aligned} \mathbf{p} &= \cos \boldsymbol{\theta} \\ \mathbf{q} &= \boldsymbol{\phi}. \end{aligned} \tag{3.42}$$

According to Langer's initial definitions, the total energy is a function of N coordinates and N canonically conjugate momenta [8]. However, even if the energy is only a function of the coordinates, the equipartition theorem holds [12]. The use of variables such as the ones defined in Eq. (3.42) ensures the Jacobian

$$J_i = \det \frac{\partial(m_{ix}, m_{iy}, m_{iz})}{\partial(p_i, q_i, 1)} = dp_i dq_i \tag{3.43}$$

is not a function of (θ_i, ϕ_i) [21]. In this thesis however, the Hessian matrix is constructed in spherical coordinates. This choice is simply a matter of convenience, as analytical expressions of the energy derivatives become heavy when using the change of variable $p = \cos\theta$. The choice of spherical coordinates requires corrections in the Hessian to take into account the spherical Jacobian, which we give in what follows. We define the spherical Hessian as

$$H = \begin{pmatrix} H_{\theta\theta} & H_{\theta\phi} \\ H_{\phi\theta} & H_{\phi\phi} \end{pmatrix}, \tag{3.44}$$

in which

$$
\begin{aligned}
H_{\theta_i\theta_j} &= \frac{\partial^2 E}{\partial\theta_i\partial\theta_j},\\
H_{\theta_i\phi_j} &= \frac{1}{\sin\theta_j}\frac{\partial^2 E}{\partial\theta_i\partial\phi_j},\\
H_{\phi_i\theta_j} &= \frac{1}{\sin\theta_i}\frac{\partial^2 E}{\partial\phi_i\partial\theta_j},\\
H_{\phi_i\phi_j} &= \frac{1}{\sin\theta_i\sin\theta_j}\frac{\partial^2 E}{\partial\phi_i\partial\phi_j}.
\end{aligned} \tag{3.45}
$$

Even though the total Hessian remains symmetric, it is necessary to remain cautious with the introduction of a DMI contribution as it makes the $H_{\theta\phi}$ submatrix non-symmetric. Therefore in general, $H_{\phi\theta} = H_{\theta\phi}^T \neq H_{\theta\phi}$, contrary to the way it was treated in Ref. [9].

3.3.2 Determination of the Dynamical Prefactor λ_+

The dynamical prefactor takes into account the dynamics of the system at the saddle point and is derived from the set deterministic Landau-Lifshitz-Gilbert (LLG) equations associated with each spin $\mathbf{m}_i$, $i = 1\ldots N$[9],

$$
\frac{d\mathbf{m}_i}{dt} = -\Big[g'\mathbf{m}_i \times \frac{\partial E}{\partial \mathbf{m}_i} + h'\big(\mathbf{m}_i \times \frac{\partial E}{\partial \mathbf{m}_i}\big) \times \mathbf{m}_i\Big], \tag{3.46}
$$

where

$$
g' = \frac{\gamma_e}{(1+\alpha^2)M_s} \tag{3.47}
$$

corresponds to the gyromagnetic ratio γ_e modified by a dimensionless damping factor $\alpha = \eta\gamma_e M_s$, in which M_s is the saturation magnetization and η is a damping parameter characterizing the coupling to the heat bath, and

$$
h' = \alpha g'. \tag{3.48}
$$

It follows that the first term on the RHS of Eq. (3.46) is the Larmor equation describing the precession of the magnetization vector $\mathbf{m}_i$, and the second term is an alignment term whose effect is measured by h'. The following derivation was adapted to spherical coordinates from [12]. We begin by reformulating Eq. (3.46) within the local spherical basis $(\mathbf{e}_r, \mathbf{e}_\theta, \mathbf{e}_\phi)$ in the following way:

$$
\begin{aligned}
\frac{d\mathbf{m}_i}{dt} = \frac{d\mathbf{e}_r}{dt} &= \frac{\partial\mathbf{e}_r}{\partial r_i}\dot{r}_i + \frac{\partial\mathbf{e}_r}{\partial\theta_i}\dot{\theta}_i + \frac{\partial\mathbf{e}_r}{\partial\phi_i}\dot{\phi}_i,\\
&= \dot{\theta}_i\mathbf{e}_\theta + \sin\theta_i\dot{\phi}_i\mathbf{e}_\phi,
\end{aligned} \tag{3.49}
$$

$$\begin{aligned}\mathbf{m}_i \times \frac{\partial E}{\partial \mathbf{m}_i} &= \mathbf{e}_r \times \left(\frac{\partial E}{\partial \theta_i}\mathbf{e}_\theta + \frac{1}{\sin\theta_i}\frac{\partial E}{\partial \phi_i}\mathbf{e}_\phi\right), \\ &= \frac{\partial E}{\partial \theta_i}\mathbf{e}_\phi - \frac{1}{\sin\theta_i}\frac{\partial E}{\partial \phi_i}\mathbf{e}_\theta,\end{aligned} \tag{3.50}$$

$$\begin{aligned}\mathbf{m}_i \times \frac{\partial E}{\partial \mathbf{m}_i} \times \mathbf{m}_i &= \left(\frac{\partial E}{\partial \theta_i}\mathbf{e}_\phi - \frac{1}{\sin\theta_i}\frac{\partial E}{\partial \phi_i}\mathbf{e}_\theta\right) \times \mathbf{e}_r, \\ &= \frac{\partial E}{\partial \theta_i}\mathbf{e}_\theta + \frac{1}{\sin\theta_i}\frac{\partial E}{\partial \phi_i}\mathbf{e}_\phi.\end{aligned} \tag{3.51}$$

By inserting Eqs. (3.49), (3.50), and (3.51) into Eq. (3.46), and equating the terms along $\mathbf{e}_\theta$ and $\mathbf{e}_\phi$ respectively, we obtain the following set of differential equations:

$$\begin{aligned}\frac{d\theta_i}{dt} &= \frac{g'}{\sin\theta_i}\frac{\partial E}{\partial \phi_i} - h'\frac{\partial E}{\partial \theta_i}, \\ \frac{d\phi_i}{dt} &= \frac{-g'}{\sin\theta_i}\frac{\partial E}{\partial \theta_i} - \frac{h'}{\sin^2\theta_i}\frac{\partial E}{\partial \phi_i}.\end{aligned} \tag{3.52}$$

The next step consists in approximating the energy close to the saddle point $(\tilde{\boldsymbol{\theta}}, \tilde{\boldsymbol{\phi}}) = (\tilde{\theta_1}, \ldots \tilde{\theta_N}, \tilde{\phi_1}, \ldots \tilde{\phi_N})$ as a Taylor series truncated to the second order term,

$$\begin{aligned}E(\theta_1, \ldots \theta_N, \phi_1, \ldots \phi_N) =& E^S(\tilde{\theta_1}, \ldots \tilde{\theta_N}, \tilde{\phi_1}, \ldots \tilde{\phi_N}) \\ &+ \frac{1}{2}\sum_{i,j}\Big\{H^S_{\theta\theta_{ij}}(\theta_i - \tilde{\theta}_i)(\theta_j - \tilde{\theta}_j) + H^S_{\phi\phi_{ij}}(\phi_i - \tilde{\phi}_i)(\phi_j - \tilde{\phi}_j) \\ &+ H^S_{\theta\phi_{ij}}(\theta_i - \tilde{\theta}_i)(\phi_j - \tilde{\phi}_j) + H^S_{\phi\theta_{ij}}(\phi_i - \tilde{\phi}_i)(\theta_j - \tilde{\theta}_j)\Big\},\end{aligned} \tag{3.53}$$

where E^S is the energy at S and $H^S_{\theta\theta}$, $H^S_{\phi\phi}$, $H^S_{\theta\phi}$, $H^S_{\phi\theta}$ are the submatrices in the spherical Hessian defined in Eq. (3.45) and evaluated at S. Taking the derivative of Eq. (3.53) in θ_i and ϕ_i then yields

$$\begin{aligned}\frac{\partial E}{\partial \theta_i} &= \sum_j \Big(H^S_{\theta\theta_{ij}}(\theta_j - \tilde{\theta}_j) + \frac{1}{2}(H^S_{\theta\phi_{ij}} + H^S_{\phi\theta_{ji}})(\phi_j - \tilde{\phi}_j)\Big), \\ \frac{\partial E}{\partial \phi_i} &= \sum_j \Big(H^S_{\phi\phi_{ij}}(\phi_j - \tilde{\phi}_j) + \frac{1}{2}(H^S_{\theta\phi_{ji}} + H^S_{\phi\theta_{ij}})(\theta_j - \tilde{\theta}_j)\Big).\end{aligned} \tag{3.54}$$

By noting that $H_{\phi\theta_{ji}} = H_{\theta\phi_{ij}}$ and $H_{\theta\phi_{ji}} = H_{\phi\theta_{ij}}$, the above equations reduce to

$$\begin{aligned}\frac{\partial E}{\partial \theta_i} &= \sum_j \Big(H^S_{\theta\theta_{ij}}(\theta_j - \tilde{\theta}_j) + H^S_{\theta\phi_{ij}}(\phi_j - \tilde{\phi}_j)\Big), \\ \frac{\partial E}{\partial \phi_i} &= \sum_j \Big(H^S_{\phi\phi_{ij}}(\phi_j - \tilde{\phi}_j) + H^S_{\phi\theta_{ij}}(\theta_j - \tilde{\theta}_j)\Big).\end{aligned} \tag{3.55}$$

Finally, by inserting (3.55) into (3.52) evaluated at S, and defining small deviations from the saddle point as $\boldsymbol{\Theta} = \boldsymbol{\theta} - \tilde{\boldsymbol{\theta}}$ and $\boldsymbol{\Phi} = \boldsymbol{\phi} - \tilde{\boldsymbol{\phi}}$, we obtain the following system of equations linearized about the saddle point, which we write in matrix form,

$$\begin{pmatrix} \dot{\boldsymbol{\Theta}} \\ \dot{\boldsymbol{\Phi}} \end{pmatrix} = \mathcal{T} \begin{pmatrix} \boldsymbol{\Theta} \\ \boldsymbol{\Phi} \end{pmatrix}, \tag{3.56}$$

in which the dot notation is a time derivative and

$$\mathcal{T} = \begin{pmatrix} \mathcal{T}_{\theta\theta} & \mathcal{T}_{\theta\phi} \\ \mathcal{T}_{\phi\theta} & \mathcal{T}_{\phi\phi} \end{pmatrix} \tag{3.57}$$

is the transition matrix of LLG, where

$$\begin{aligned}
\mathcal{T}_{\theta_i\theta_j} &= g' H^S_{\phi_i\theta_j} - h' H^S_{\theta_i\theta_j}, \\
\mathcal{T}_{\theta_i\phi_j} &= \sin\tilde{\theta}_j \left(g' H^S_{\phi_i\phi_j} - h' H^S_{\theta_i\phi_j} \right), \\
\mathcal{T}_{\phi_i\theta_j} &= -\frac{1}{\sin\tilde{\theta}_i} \left(g' H^S_{\theta_i\theta_j} + h' H^S_{\phi_i\theta_j} \right), \\
\mathcal{T}_{\phi_i\phi_j} &= -\frac{\sin\tilde{\theta}_j}{\sin\tilde{\theta}_i} \left(g' H^S_{\theta_i\phi_j} + h' H^S_{\phi_i\phi_j} \right),
\end{aligned} \tag{3.58}$$

define its submatrices. Similarly to the Hessian, the transition matrix is a $2N \times 2N$ matrix, but the number of positive eigenvalues it possesses equates to the order of the saddle point. The diagonalized transition matrix contains the eigenfrequencies of the system at the saddle point. In the most common configuration where the saddle point is of first order, the single positive eigenvalue of $\mathcal{T}$ is taken as the dynamical contribution to the attempt frequency, λ_+, and is interpreted as the growth rate the dynamically unstable deviation at the saddle [9]. In the case of a higher-order saddle point, it is not clear whether Langer's theory still applies. In [12], it was proposed that if several unstable modes are present at S, λ_+ should be chosen as the biggest—i.e., most positive—eigenvalue of $\mathcal{T}$, thus giving the rate of the fastest growing instability. We can note that the transition matrix is not symmetric and can in principle admit complex eigenvalues and eigenvectors. Although we have only encountered real eigenvalues in this thesis, one example of a complex λ_+ can be found in Ref. [12] in the case of the antiferromagnetic reversal of two ferromagnetic particles. In that case, λ_+ should be taken as the largest real part amongst the eigenvalues. Rewriting Eq. (3.57) in the eigenbasis yields

$$\begin{pmatrix} \dot{\boldsymbol{\Theta}}' \\ \dot{\boldsymbol{\Phi}}' \end{pmatrix} = \mathcal{T}' \begin{pmatrix} \boldsymbol{\Theta}' \\ \boldsymbol{\Phi}' \end{pmatrix}, \tag{3.59}$$

where $(\boldsymbol{\Theta}', \boldsymbol{\Phi}')$ are the eigenbasis coordinates and $\mathcal{T}'$ is the diagonalized transition matrix. In particular, the unstable mode evolves as

$$\big(\Theta'_+(t), \Phi'_+(t)\big) = e^{\lambda_+ t}\big(\Theta'_+(t=0), \Phi'_+(t=0)\big). \tag{3.60}$$

If λ_+ is real, the instability grows exponentially. On the other hand, if λ_+ is a complex number, $\lambda_+ = \sigma + i\omega$ with $\sigma, \omega \in \mathbb{R}$, $i^2 = -1$, and $\sigma > 0$, so

$$\big(\Theta'_+(t), \Phi'_+(t)\big) = e^{\sigma t} e^{i\omega t}\big(\Theta'_+(t=0), \Phi'_+(t=0)\big), \tag{3.61}$$

in which $e^{i\omega t}$ is a phasor. In that case, the time evolution of the instability is a sinusoid inside a growing exponential envelope. In all cases, the timescale of the transition is set by the damping and the characteristic precession time, $M_s/(J_1\gamma_e)$.

3.4 The Geodesic Nudged Elastic Band Method

The nudged elastic band method (NEB, or NEBM) [22, 23] is a numerical scheme designed to compute paths of minimum energy through the (often complex) multidimensional energy landscape of a system connecting two energy minima. The NEB method has been most commonly used for transitions involving atomic rearrangements. The method consists in specifying a discrete representation of an initial path between the two minima corresponding to the initial and final states of the system for the transition of interest, and iteratively bringing the path to the nearest minimum energy path (MEP). Each discretization point along the path corresponds to a state of the complete system in the space of configurations, and is referred to as an image. Adjacent images are connected with springs that allow control over the resolution of the path and prevent the images from sliding down towards energy minima. A considerable advantage of the NEB method over similar methods of finding a MEP is that it does not require an initial guess of the reaction coordinate. Instead, it is naturally found as the path relaxes to a MEP, at which point the tangent to the path provides a good estimate of the reaction coordinate. A MEP is a path of highest statistical weight and therefore corresponds to a particular mechanism of the considered transition. As was touched upon in the previous section, there can be one or several MEP(s), each of them associated with a specific mechanism and transition rate. In particular, the NEB method can be combined with a scheme that allows a precise identification of the first-order saddle point along the path, such as the climbing image (CI) NEB, known as CI-NEB [24], and thus constitutes an important building block for a numerical implementation of reaction rate theory.

The geodesic NEB method (GNEB) is a variant of the NEB method adapted to magnetic systems which was developed by Bessarab et al. [11]. It solves potential convergence problems of the NEB method when applied to magnetic spin systems, by taking into account the curved geometry of the space of configurations. Its key features lie in the use of geodesics to calculate distances between images, and in the projection of the tangent to the path and of the force acting on the images onto the space tangent to the space of configurations. In what follows, we expose the principles of this method and provide details of its implementation.

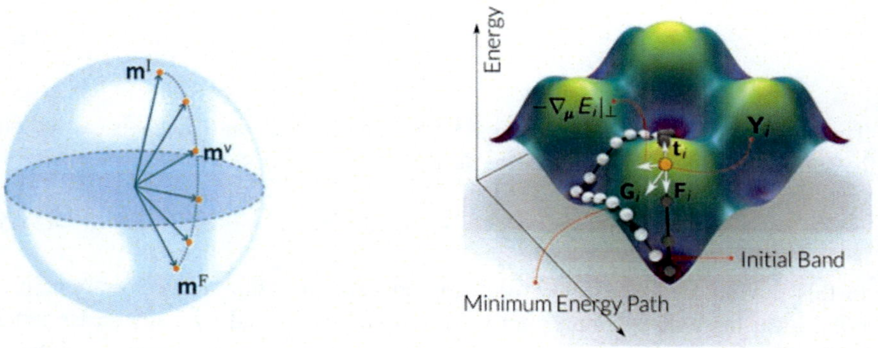

(a) (b)

Fig. 3.3 Illustration of the GNEB method. **a** Illustration of the GNEB method in the space of configurations for a single spin system $\mathbf{m}$. In this case, the Riemannian manifold is the unit sphere S^2. The path between the initial state $\mathbf{m}^I$ and the final state $\mathbf{m}^F$ is discretized into a series of images $\{\mathbf{m}^\nu\}$, $\nu = 1 \dots Q$, marked as orange dots. **b** Illustration of the GNEB method on a two-dimensional energy surface [25]. The elastic band is shown as a chain of beads representing images and connected by a black string which representing springs. The initial path is shown in black and passes through a second-order saddle point (here, a local maximum). After the CI-GNEB procedure, the path relaxes to the MEP shown in white, which passes through two first-order saddle points separated by a local minimum

3.4.1 Projections onto the Tangent Space to the Manifold

The rest of this section is based on [11]. As we have assumed that the amplitude of the magnetic vectors remains fixed, each magnetic vector $\mathbf{m}_i$ thus lives on the unit sphere associated with site i, S_i^2, with its orientation specified by two coordinates (Fig. 3.3a). For instance, one may choose the spherical angles (θ, ϕ). The configuration space of a system of N spins is the $2N$-dimensional Riemannian manifold $\mathscr{R}$ obtained from a direct product of the N two-dimensional unit spheres S_i^2,

$$\mathscr{R} = \prod_{i=1}^{N} S_i^2. \tag{3.62}$$

In the GNEB scheme, the local tangent to path, as well as the total force acting on each image, are projected onto the tangent space to the manifold. We define the tangent space to the unit sphere, S_i^2, at $\mathbf{m}_i$, as the span of all vectors $\mathbf{x}$ normal to $\mathbf{m}_i$, that is,

$$\mathcal{T}_{\mathbf{m}_i} S_i^2 = \{\mathbf{x} : \mathbf{x} \cdot \mathbf{m}_i = 0\}. \tag{3.63}$$

The tangent space to the $2N$-dimensional Riemannian manifold $\mathscr{R}$ is simply defined as the product of the tangent spaces to each sphere,

$$\mathcal{T}_{\mathbf{m}_i}\mathscr{R} = \prod_{i=1}^{N} \mathcal{T}_{\mathbf{m}_i} S_i^2. \tag{3.64}$$

The projection of a three-dimensional Cartesian vector $\mathbf{a}_i$ onto $\mathcal{T}_{\mathbf{m}_i}\mathscr{R}$ is carried out by substracting its component along $\mathbf{m}_i$,

$$\mathbf{a}_{i,\mathcal{T}} = \mathbf{a}_i - (\mathbf{a}_i \cdot \mathbf{m}_i)\mathbf{m}_i. \tag{3.65}$$

In this section, we choose the following convention: while all vectors remain in bold, local three dimensional vectors $\mathbf{a}$ will be represented by a lowercase letter, while $3N$-dimensional vectors are written as the corresponding uppercase letter as $\mathbf{A} = (\mathbf{a}_1, \mathbf{a}_2, \ldots \mathbf{a}_N)$. We can now define the projection of a $3N$-dimensional vector $\mathbf{A}$ onto the tangent space to the $\mathscr{R}$-manifold as

$$\mathcal{P}_{\mathcal{T}}\mathbf{A} = (\mathbf{a}_{1,\mathcal{T}}, \mathbf{a}_{2,\mathcal{T}}, \ldots \mathbf{a}_{N,\mathcal{T}}). \tag{3.66}$$

3.4.2 The GNEB Method

We construct a chain of Q images, $[\mathbf{M}^1, \mathbf{M}^2 \ldots \mathbf{M}^Q]$ (Fig. 3.3a), where each image $\mathbf{M}^\nu = (\mathbf{m}_1^\nu, \mathbf{m}_2^\nu \ldots \mathbf{m}_N^\nu)$, with $\nu = 1 \ldots Q$, is a $3N$-dimensional Cartesian vector. The chain of images constitutes the elastic band. The first and last images, $\mathbf{M}^1 = \mathbf{M}^I$ and $\mathbf{M}^Q = \mathbf{M}^F$, are chosen as energy minima corresponding to the initial and final states of the transition, and remain fixed. The remaining images are iteratively brought to a MEP through minimization of the total GNEB force acting on them (Fig. 3.3b). That force is the sum of the negative energy gradient perpendicular to the path, i.e., the true force, and a spring force acting along the path. This requires an estimation of the tangent to the path at each image.

The tangent to the path at image ν, $\boldsymbol{\tau}^\nu$, is evaluated as detailed in Sect. 3.4.3, and projected onto the tangent space according to Eq. (3.66),

$$\boldsymbol{\tau}_{\mathcal{T}}^\nu = \frac{\mathcal{P}_{\mathcal{T}}\boldsymbol{\tau}^\nu}{|\,\mathcal{P}_{\mathcal{T}}\boldsymbol{\tau}^\nu\,|}. \tag{3.67}$$

The true force acting on image ν originates from the transverse (perpendicular) component of the negative gradient, and is obtained by subtracting the component of the gradient along the tangent to the path,

$$\nabla E(\mathbf{M}^\nu)_\perp = \nabla E(\mathbf{M}^\nu) - \left(\nabla E(\mathbf{M}^\nu \cdot \boldsymbol{\tau}_{\mathcal{T}}^\nu\right)\boldsymbol{\tau}_{\mathcal{T}}^\nu. \tag{3.68}$$

The spring force, on the other hand, only has a component along the path tangent and its magnitude depends on the geodesic distances between adjacent images, that is,

$$\mathbf{F}^{\nu}_{S\|} = \kappa^{\nu}\big[L(\mathbf{M}^{\nu+1}, \mathbf{M}^{\nu}) - L(\mathbf{M}^{\nu}, \mathbf{M}^{\nu-1})\big]\boldsymbol{\tau}^{\nu}_{\mathcal{T}}. \quad (3.69)$$

In the previous expression, κ^{ν} is the spring constant at image ν, which can be chosen to be constant at all images, or it may be adjusted to gain resolution in critical areas, typically around a saddle point. We give details on the variable spring scheme [24] later in this section. $L(\mathbf{M}^{\nu}, \mathbf{M}^{\mu})$ is the geodesic distance between two images ν and μ on the $\mathscr{R}$-manifold. It is defined as

$$L(\mathbf{M}^{\nu}, \mathbf{M}^{\mu}) = \sqrt{(l_1^{\nu,\mu})^2 + (l_2^{\nu,\mu})^2 + \cdots + (l_N^{\nu,\mu})^2}, \quad (3.70)$$

where the $l_i^{\nu,\mu}$ is the local geodesic distance between ν and μ on the unit sphere S_i^2, which we evaluate as

$$l_i^{\nu,\mu} = 2\arcsin\sqrt{\frac{1 - \mathbf{m}^{\nu}\cdot\mathbf{m}^{\mu}}{2}}. \quad (3.71)$$

This projection of the true force and the spring force using the tangent to the path is referred to as "nudging". Finally, the total GNEB force acting on image ν needs to be projected onto the tangent space. In the end, the force applied to the images is expressed as

$$\mathbf{F}^{\nu}_{\text{GNEB}} = \mathcal{P}_{\mathcal{T}}\big(-\nabla E(\mathbf{M}^{\nu})_{\perp} + \mathbf{F}^{\nu}_{S\|}\big). \quad (3.72)$$

The projection of the tangent to the path onto the tangent space to the curved manifold allows a decoupling of the spring force and the true force, so that they do not interfere with each other. If this procedure is not carried out, the control over the distribution of images is lost and the scheme becomes sensitive to the value of the spring constant [11]. The path will usually converge to the MEP closest to the initial path. If several transition mechanisms exist, several runs of the GNEB scheme with different initializations of the path must to be carried out, until all possible MEPs are identified. An example run of the GNEB scheme is shown in Fig. 3.4 for a single spin reversal with a hard-axis along z. The path is initialized out of plane with some random noise added to the initial distribution of the images. The scheme iteratively brings the path to the xy-plane and uniformly redistributes the images.

3.4.3 Evaluation of the Tangent to the Path

An appropriate definition of the tangent to the path at image ν, $\boldsymbol{\tau}^{\nu}$, is one key aspect to the stability of the method. We used the definition from [11],

$$\boldsymbol{\tau}^{\nu} = \begin{cases} \boldsymbol{\tau}^{\nu}_{+}, & \text{if } E^{\nu+1} > E^{\nu} > E^{\nu-1}, \\ \boldsymbol{\tau}^{\nu}_{-}, & \text{if } E^{\nu+1} < E^{\nu} < E^{\nu-1}, \end{cases} \quad (3.73)$$

where $\boldsymbol{\tau}_+^\nu = \mathbf{M}^{\nu+1} - \mathbf{M}^\nu$ and $\boldsymbol{\tau}_-^\nu = \mathbf{M}^\nu - \mathbf{M}^{\nu-1}$, and $E^\nu = E(\mathbf{M}^\nu)$. If the image is close to a maximum or a minimum in the energy,

$$\boldsymbol{\tau}^\nu = \begin{cases} \boldsymbol{\tau}_+^\nu \Delta E_{\text{max}}^\nu + \boldsymbol{\tau}_-^\nu \Delta E_{\text{min}}^\nu & \text{if } E^{\nu+1} > E^{\nu-1}, \\ \boldsymbol{\tau}_+^\nu \Delta E_{\text{min}}^\nu + \boldsymbol{\tau}_-^\nu \Delta E_{\text{max}}^\nu & \text{if } E^{\nu+1} < E^{\nu-1}, \end{cases} \tag{3.74}$$

where $\Delta E_{\text{max}}^\nu = \max\big(|E^{\nu+1} - E^\nu|, |E^{\nu-1} - E^\nu|\big)$ and $\Delta E_{\text{min}}^\nu = \min\big(|E^{\nu+1} - E^\nu|, |E^{\nu-1} - E^\nu|\big)$. Equation (3.74) provides a smooth transition between the forward and backward definitions, $\boldsymbol{\tau}_+^\nu$ and $\boldsymbol{\tau}_-^\nu$. The above definition of the tangent to the path according to Eqs. (3.73) and (3.74) enhances the stability of the method by preventing the formation of kinks in the path [26].

3.4.4 *Initialization of the Path Along Geodesics*

The initialization of the GNEB scheme requires an initial path between the initial states to be generated. If a good guess cannot be made, a solution is to evenly distribute the images along the geodesic path connecting the first and last image. We have followed the procedure described in [11], which makes use of Rodrigues' rotation formula to generate the orientation of the local magnetic vector $\mathbf{m}_i^\nu$ at image ν as

$$\mathbf{m}_i^\nu = \mathbf{m}_i^F \cos\omega_i^\nu + (\mathbf{k}_i \times \mathbf{m}_i^I) \sin\omega_i^\nu, \tag{3.75}$$

where ω_i^ν is the rotation angle and $\mathbf{k}_i$ is the rotation vector. The angle of rotation ω_i^ν is a fraction of the total angle ω_i between $\mathbf{m}_i^I$ and $\mathbf{m}_i^F$,

$$\omega_i^\nu = \frac{(\nu - 1)\omega_i}{Q - 1}, \tag{3.76}$$

and $\omega_i = \arccos(\mathbf{m}_i^I \cdot \mathbf{m}_i^F)$. Finally, the normalized axis of rotation is defined by

$$\mathbf{k}_i = \frac{\mathbf{m}_i^I \times \mathbf{m}_i^F}{|\,\mathbf{m}_i^I \times \mathbf{m}_i^F\,|}. \tag{3.77}$$

3.4.5 *Search for the First-Order Saddle Point: The CI-GNEB Scheme*

If one is only looking to estimate the energy barrier, this can in principle be done by interpolating the energy along the relaxed path. However, the use of Langer's theory to calculate a rate prefactor, like any modern form of reaction rate theory, requires a precise identification of the saddle point along the MEP. This can be achieved by a climbing image (CI) scheme [24]. After the regular GNEB scheme has relaxed down to a certain tolerance on the GNEB force, the highest energy image becomes a climbing image. The spring force on this particular image is dropped, and instead the CI is made to move uphill in energy along the reaction coordinate, which is defined by the tangent to the path, and downhill in energy along all other degrees of freedom. For that purpose, the force acting on the CI takes the form,

$$\mathbf{F}^{\nu_{CI}}_{GNEB} = \mathcal{P}_{\mathcal{T}}(-\nabla E(\mathbf{M}^{\nu_{CI}}) + 2(\nabla E(\mathbf{M}^{\nu_{CI}}) \cdot \tau^{\nu_{CI}}_{\mathcal{T}})\tau^{\nu_{CI}}_{\mathcal{T}}). \tag{3.78}$$

By minimizing the force acting on it, the CI will, in most cases, relax to the highest energy first-order saddle point on the path. If an energy interpolation reveals the existence of multiple saddle points along the path, two or more climbing images may be specified. During the CI-GNEB procedure, the rest of the images still need to be displaced according to Eq. (3.72). Like previously, the procedure ends once the force on *all* images is relaxed below a set tolerance. We can verify that we have indeed found a first-order saddle point by checking that the Hessian evaluated at that point possesses only a single negative eigenvalue.

3.4.6 Variable Spring Constants

The spring force in Eq. (3.69) that controls the distribution of images along the path may be adjusted as to increase the resolution of the path in certain areas of interest. Typically, a better resolution around the saddle point will increase the accuracy of the path tangent and provide a better estimate of the reaction coordinate for the climbing image to follow. Because the true force and the spring force are decoupled, the use of different springs between different pairs of images does not affect the convergence to the MEP, as long as the number of images is sufficient. We have used the scheme proposed in [24] which results in lower energy images being connected by weaker springs, as follows,

$$\kappa^{\nu} = \begin{cases} \kappa_{\max} - \Delta\kappa \left(\dfrac{E_{\max} - E^{\nu}_{\max}}{E_{\max} - E^{\text{ref}}} \right) & \text{if } E^{\nu}_{\max} > E^{\text{ref}}, \\ \kappa_{\max} - \Delta\kappa & \text{if } E^{\nu}_{\max} \leq E^{\text{ref}}. \end{cases} \tag{3.79}$$

κ^{ν} is the spring constant associated with spring ν (where $\nu = 2 \dots Q$) connecting images $\nu - 1$ and ν, $E^{\nu}_{\max} = \max(E^{\nu-1}, E^{\nu})$ is the highest energy of the two images connected by spring ν, and $E_{\max}$ is the maximum value of E^{ν} amongst all images. E^{ref} is a reference value for the energy, which we set as the highest energy of the initial and final states, $E^{\text{ref}} = \max(E^{I}, E^{F})$ as suggested in [24]. This results in a maximum value of the spring constant of $\kappa_{\max}$ between highest energy images, and a minimum value of $\kappa_{\max} - \Delta\kappa$ for images with energy E^{ref} or lower.

We have found that variable springs can sometimes improve the convergence of the method, in particular in systems where the energy landscape is very complex,

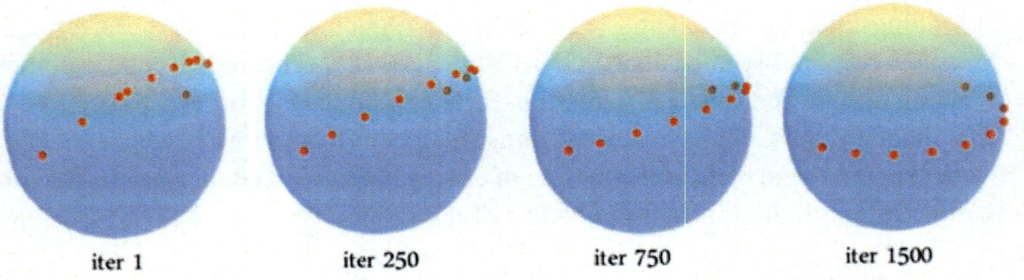

Fig. 3.4 Illustration of the GNEB scheme for a single spin **m** undergoing a magnetization reversal from $m_x = -1$ to $m_x = 1$ with easy-axis anisotropy along $\mathbf{e}_x$, and hard-axis anisotropy along $\mathbf{e}_z$. The initialization is done out-of-plane with a random noise added to the distribution of the images and the scheme iteratively brings the images in-plane and redistributes them uniformly along the path

with many local extrema. In this work, this mainly concerns the paths to collapse of topological defects in exchange-frustrated systems which will be presented in Chap. 6.

3.5 Implementation of Langer's Theory on an Atomistic Spin Lattice

3.5.1 Atomistic Simulations and Search for Local Minima

The following methods were implemented by the author as an atomistic code in C++ language over the course of this thesis. The calculations were carried out in Cartesian coordinates, with the exception of the rate prefactor. All quantities are dimensionless, in units of the isotropic exchange coupling constant between first neighbours, $J_1 = J_{\text{ex}}$, as defined in Sect. 2.1.4. The atomistic Hamiltonian on the two-dimensional square lattice is calculated from Eq. (2.1). The reduced local effective field acting on spin $\mathbf{m}_i$ at site i is

$$\mathbf{b}^i_{\text{eff}} = -\frac{1}{J_1}\frac{\partial E}{\partial \mathbf{m}_i}. \tag{3.80}$$

We relax local energy minima by the simple iterative energy minimization procedure described in Sect. 11.2 of Ref. [27]. The system is initialized close to a local energy minimum of interest. For instance, we initialize a central square of reversed spin on a ferromagnetic background in order to relax a skyrmion. After that, a normalized effective field at spin i,

$$\hat{\mathbf{b}}^i_{\text{eff}} = \mathbf{b}^i_{\text{eff}} / \mid \mathbf{b}^i_{\text{eff}} \mid,$$

is calculated according to Eq. (3.80), and $\mathbf{m}_i$ is aligned onto the local field by simply setting $\mathbf{m}_i = \hat{\mathbf{b}}^i_{\text{eff}}$. We then move on to another spin and proceed analogously. The iterations end when the largest component of $\mathbf{m}_i \times \hat{\mathbf{b}}^i_{\text{eff}}$ falls below a set tolerance ϵ. In general we set $\epsilon \in [10^{-8} - 10^{-15}]$. This method converges to the minimum that is closest to the initialization state, and is therefore well adapted to the search for metastable states. While it may fail for very simple cases (eg: a single macrospin), it works well for $N-$spin systems ($N > 1$) and yields the same results as the integration of the LLG equation (Eq. (3.46)) at zero temperature.

3.5.2 Implementation of GNEB

3.5.2.1 Overview of the GNEB Algorithm

We give an overview of the GNEB and CI-GNEB algorithm:

- The first and last images, $\mathbf{M}^I$ and $\mathbf{M}^F$, are initialized as energy minima. These are usually previously generated via the method described in Sect. 3.5.1, or they may also be relaxed through integration of the LLG equation (3.46), or via the GNEB scheme [11].
- An initial path $[\mathbf{M}^2, \mathbf{M}^3 \ldots \mathbf{M}^{Q-1}]$ between $\mathbf{M}^I$ and $\mathbf{M}^F$ is generated, either along geodesics as described in Sect. 3.4.4, or it is manually implemented as part of an initial guess.
- The GNEB scheme begins and runs until all Cartesian components of the local force on each spin at each image falls below a set tolerance F_{tol}.
- The highest energy image is selected as the climbing image and the CI-GNEB scheme runs until the force falls below the set tolerance.
- The energy between the images is interpolated by a piecewise cubic interpolation, as described in [11].

3.5.2.2 Choice of Solver

Although one could choose to integrate a precessionless version of the LLG equation at each timestep, in our case the force minimization during the (CI)-GNEB procedure was done as suggested in [11], by simply integrating Newton's equations for the motion of a fictitious point-mass on the unit sphere (Fig. 3.4), where the total force is the GNEB force, i.e.,

$$\mathbf{F}^{\nu}_{\text{GNEB}} = m\Gamma^{\nu}. \tag{3.81}$$

In the previous expression, m is an arbitrary mass that controls the inertia of the images, which we set to $m = 1$, and $\Gamma^{\nu} = d^2\mathbf{M}^{\nu}/dt^2 = (\boldsymbol{\gamma}^{\nu}_1, \boldsymbol{\gamma}^{\nu}_2 \ldots \boldsymbol{\gamma}^{\nu}_N)$ is the acceleration of image ν. The value of the spring constant $\kappa_{\max}$ is not critical in GNEB [11], but large values tend to make the path longer in certain cases, so we set $\kappa_{\max} = 1$

and $\Delta\kappa = 0.1$. Depending on the problem at hand, we typically use $Q = 8 - 15$ images. The tolerance on each Cartesian component of the GNEB force at each image and each spin is set around $F_{\text{tol}} = 10^{-10} - 10^{-12}$. Although Ref. [11] recommends the use of spherical coordinates so that the constraints on the magnetic vectors are automatically fulfilled, in practice, because the spherical coordinate system has singularities at the poles of the sphere (i.e., the azimuth ϕ is not defined for $\theta = 0$ or π), a special treatment is required in these areas. Therefore, we use Cartesian coordinates instead, and re-normalize the magnetic vectors at each iteration. We use the velocity Verlet integrator [28], whose global error is of order 2, with an integration step $\Delta t = 0.01$.

The convergence is aided by the velocity projection optimization (VPO) procedure [11]. It consists in damping the velocity at each time-step by keeping only its component along the force. That is, unless the velocity is pointing in a direction opposite to that of the force, in which case it is set to zero. A similar method was used for NEB calculations in unconstrained Euclidian space [23]. We define the velocity at image ν,

$$\mathbf{V}^{\nu} = \frac{d\mathbf{M}^{\nu}}{dt} = (\mathbf{v}_1^{\nu}, \mathbf{v}_2^{\nu} \ldots \mathbf{v}_N^{\nu}). \tag{3.82}$$

Let

$$\mathscr{F} = [\mathbf{F}^1_{\text{GNEB}}, \mathbf{F}^2_{\text{GNEB}} \ldots \mathbf{F}^Q_{\text{GNEB}}] \tag{3.83}$$

be a $3N \times Q$-dimensional vector of the force, and

$$\mathscr{V} = [\mathbf{V}^1, \mathbf{V}^2 \ldots \mathbf{V}^Q] \tag{3.84}$$

a $3N \times Q$-dimensional vector of the velocity. For convenience, we define $a = \mathscr{V} \cdot \mathscr{F}/Q$ and $b = \mathscr{F} \cdot \mathscr{F}/Q$ as vector projections in $3N \times Q$-dimensional space. At each timestep, the velocity is revised according to [11, 29],

$$\mathscr{V} = \begin{cases} a\mathscr{F}/b & \text{if } a > 0, \\ 0 & \text{if } a \leq 0, \end{cases} \tag{3.85}$$

and used in the next iteration. As a result, if the force keeps pointing in the same direction, the system accelerates and approaches the minimum faster. Then, if the system overshoots the minimum, the velocity is set to zero.

3.5.2.3 Force Minimization Algorithm

The aim of the GNEB method is to relax the initial path onto a MEP by minimizing the GNEB force acting on the images. One proceeds as follows. First, the scheme is initialized as described below. At each image $\nu = 1 \ldots Q$:

- the energy $E^{\nu} = E(\mathbf{M}^{\nu})$ is calculated;

- the tangent to the path is calculated as described in Sect. 3.4.3, and projected onto the tangent space according to Eq. (3.67);
- the total force $\mathbf{F}^{\nu}_{\text{GNEB}}$ is calculated and projected onto the tangent space according to Eq. (3.72);
- the acceleration is initialized as $\Gamma^{\nu} = \mathbf{F}^{\nu}_{\text{GNEB}}/m$;
- the velocity $\mathbf{V}^{\nu}$ is set to zero.

This completes the initialization. After that, the following procedure is iterated until the set tolerance on the components of the GNEB force has been reached:

- the position of each image $\mathbf{M}^{\nu}$ is updated according to

$$\mathbf{M}^{\nu}(t+dt) = \mathbf{M}^{\nu}(t) + \mathbf{V}^{\nu}(t)dt + \frac{1}{2}\Gamma^{\nu}(t)dt^2, \tag{3.86}$$

  and renormalized;
- then, at each image:
  - the new energy $E^{\nu}(t+dt) = E(\mathbf{M}^{\nu}(t+dt))$ is calculated;
  - the new projected tangent to the path $\boldsymbol{\tau}^{\nu}_{T}(t+dt)$ is calculated;
  - the new projected GNEB force $\mathbf{F}^{\nu}_{\text{GNEB}}(t+dt)$ is calculated and its local components are compared to the set tolerance;
  - the acceleration is updated as $\Gamma^{\nu}(t+dt) = \mathbf{F}^{\nu}_{\text{GNEB}}(t+dt)/m$;
  - the velocity is updated according to

$$\mathbf{V}^{\nu}(t+dt) = \mathbf{V}^{\nu}(t) + \frac{1}{2}\Big(\Gamma^{\nu}(t+dt) + \Gamma^{\nu}(t)\Big)dt; \tag{3.87}$$

- the velocity $\mathbf{V}^{\nu}(t+dt)$ is revised at each image according to the VPO procedure.

3.5.3 Computation of the Rate Prefactor

Once the CI-GNEB scheme has relaxed to a low tolerance, we have in principle identified the configuration of the (highest) first-order saddle point S along the reaction coordinate to high accuracy. This gives us a straightforward value for the internal energy barrier as $\Delta E = E_S - E_A$. One can now compute the pre-exponential factor in Langer's theory, as given in Eq. (3.31). Since the space of configurations is $2N$-dimensional, we found the use of spherical coordinates to be a better choice at this point in the computation. We thus begin by converting the positions of states A and S into spherical angles (θ, ϕ). If at least one spin $\mathbf{m}_i$ lies in the vicinity of the poles of the sphere, i.e., $\mid \sin\theta_i \mid < \varepsilon$, where ε is chosen as $\varepsilon \approx 10^{-8}$, the whole system is rotated about the y-axis by an arbitrary angle, until there are no spins close to the poles. This includes a rotation of the external magnetic field, the anisotropy axis, and the Dzyaloshinskii vector, as well as all the magnetic vectors.

After that, we can construct the submatrices in the Hessian matrix evaluated at A and S, as previously defined in Eqs. (3.44) and (3.45). In order to keep numerical noise to a minimum, these were implemented with analytical expressions for the second derivatives of the different terms of the Hamiltonian. The Hessian is then diagonalized using the eigensystems functions for real symmetric matrices from the GNU scientific library [30]. The library uses the symmetric bidiagonalization and QR reduction method. The $2N$ eigenvalues $\{\lambda_i\}$ are then sorted by increasing value, and we also store the corresponding eigenvectors $\{\chi_i\}$. It is a good idea to check that there is indeed only one negative eigenvalue at S, a sign that we have found a first-order saddle point.

Next, the determination of λ_+ is carried out. We use the submatrices of the Hessian to construct the transition matrix $\mathcal{T}$, as defined in Eqs. (3.57) and (3.58). To extract λ_+, we can take advantage of the fact that, as long as longe-range dipolar couplings are not included in the Hamiltonian, $\mathcal{T}$ is a sparse matrix. For that purpose, we use the Spectra library [31] built on top of the open source linear algebra library Eigen [32].

Finally, one has to test for possible zero-eigenvalue(s) in the Hessian. If some eigenvalues are close to 0, their contribution to the prefactor needs to be handled separately, as described in Sect. 3.2.2.1. Otherwise, the ratio of eigenvalues can be calculated, and we obtain the rate prefactor of the transition.

3.5.4 Numerical Tests

3.5.4.1 One Dimensional Domain Wall Profile

We first of all relaxed a one dimensional domain wall, and checked the profile against the theoretical expression (Eq. 2.32). Both the atomistic and theoretical profiles are plotted in Fig. 3.5a and show excellent agreement.

3.5.4.2 Skyrmion Profile

The relaxed skyrmion profile was compared with the output of the micromagnetic simulation program MuMax3 [33]. We give the following reduced parameters, as defined in Sect. 2.1.4: $(d, k, b) = (0.36, 0.4, 0.0)$, which allow the existence of small isolated skyrmions at zero applied field [34]. We simulated a square lattice of 50×50 spins with open (non periodic) boundary conditions. The profile is plotted in Fig. 3.5b and agrees well with the output of MuMax3. Here, we used a continuous definition of the skyrmion radius, as given in Ref. [34], which is the intersection of the largest gradient of the interpolated skyrmion profile with the $\theta = 0$ axis. The largest gradient line is shown in blue in Fig. 3.5b and intersects the horizontal axis at $r_{\text{sk}} = 5.83\ a$.

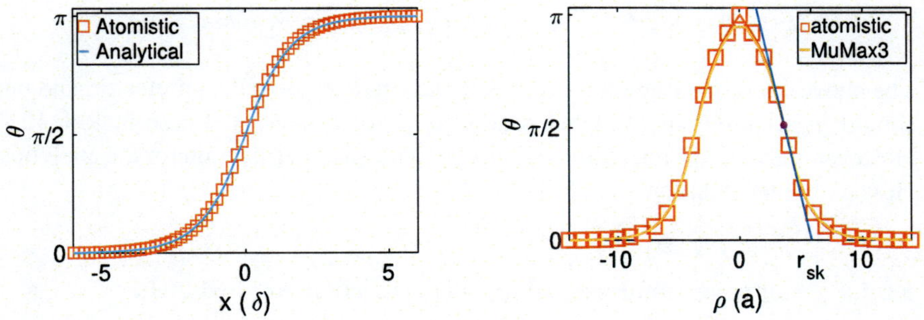

(a) (b)

Fig. 3.5 **a** Analytical micromagnetic and numerical atomistic π-domain wall profiles as a function of the spatial displacement from the wall centre, expressed in units of the characteristic wall length δ. The analytical profile was previously derived in Sect. 2.2.1 and its expression was given in Eq. (2.32). **b** Comparison of the relaxed skyrmion profiles at zero temperature and reduced parameters $(d, k, b) = (0.36, 0.4, 0.0)$, as computed by our atomistic implementation and the micromagnetic program MuMax3. The skyrmion radius is given by the intersection of the largest gradient line of the interpolated profile (shown in blue) with the horizontal $\theta = 0$ axis

3.5.4.3 Test for the Hessian Implementation

A straightforward way to test for the correct analytical implementation of the Hessian is to compare it with a numerical implementation using finite differences. Let $d\theta$ and $d\phi$ be small angular variations about a given state $(\boldsymbol{\theta}, \boldsymbol{\phi})$, with energy $E(\boldsymbol{\theta}, \boldsymbol{\phi}) = E_0$. We set, for instance, $d\theta = d\phi = 0.001$. We give the expressions for the submatrices in the Hessian, as defined in Eq. (3.45), using finite differences,

$$H_{\theta_i\theta_j} \approx \begin{cases} \frac{E(\theta_i+d\theta,\theta_j+d\theta)-E(\theta_i+d\theta,\theta_j-d\theta)-E(\theta_i-d\theta,\theta_j+d\theta)+E(\theta_i-d\theta,\theta_j-d\theta)}{4d\theta^2} & \text{if } i \neq j, \\ \frac{E(\theta_i+d\theta)-2E_0+E(\theta_i-d\theta)}{d\theta^2} & \text{if } i = j, \end{cases} \tag{3.88}$$

$$H_{\theta_i\phi_j} \approx \frac{E(\theta_i + d\theta, \phi_j + d\phi) - E(\theta_i + d\theta, \phi_j - d\phi) - E(\theta_i - d\theta, \phi_j + d\phi) + E(\theta_i - d\theta, \phi_j - d\phi)}{4d\theta \sin\theta_j d\phi}, \tag{3.89}$$

$$H_{\phi_i\theta_j} \approx \frac{E(\theta_j + d\theta, \phi_i + d\phi) - E(\theta_j + d\theta, \phi_i - d\phi) - E(\theta_j - d\theta, \phi_i + d\phi) + E(\theta_j - d\theta, \phi_i - d\phi)}{4d\theta \sin\theta_i d\phi}, \tag{3.90}$$

$$H_{\phi_i\phi_j} \approx \begin{cases} \frac{E(\phi_i+d\phi,\phi_j+d\phi)-E(\phi_i+d\phi,\phi_j-d\phi)-E(\phi_i-d\phi,\phi_j+d\phi)+E(\phi_i-d\phi,\phi_j-d\phi)}{4d\phi^2 \sin\theta_i \sin\theta_j} & \text{if } i \neq j, \\ \frac{E(\phi_i+d\phi)-2E_0+E(\phi_i-d\phi)}{d\phi^2 \sin\theta_i \sin\theta_j} & \text{if } i = j, \end{cases} \tag{3.91}$$

Note that notations of the type $E(\theta_i + d\theta, \phi_j + d\phi)$ are a compact way to write

$$E(\theta_1, \theta_2, \ldots, \theta_i + d\theta, \ldots, \theta_N, \phi_1, \phi_2, \ldots, \phi_j + d\phi, \ldots, \phi_N).$$

The above expressions are not sufficiently accurate to compute a prefactor, and one should in principle go to higher orders, as is for instance the case in Ref. [21]. However, they are sufficient to provide a good mean of verification for the analytical Hessian implementation.

3.5.4.4 Magnetization Reversal in a Perpendicular Applied Field

Next, the complete implementation of Langer's theory was tested against a simple analytical case, which we reproduced from Ref. [21]. We looked at magnetization reversal of a small magnetic square in two dimensions. The Hamiltonian contains an easy-axis anisotropy term and an external magnetic field perpendicular to the easy axis. Analytically, the system can be approximated by a single macrospin, as long as the size of the system remains small, so that the contribution from collective modes leads to a negligible error in the attempt frequency. We write the Hamiltonian of the single macrospin in spherical coordinates,

$$E = -K(\sin^2\theta \cos^2\phi + 2h \sin\theta \sin\phi), \tag{3.92}$$

where

$$h = \frac{BM_s}{2K}, \tag{3.93}$$

is the reduced perpendicular field. Note that we are using atomistic quantities as defined in Sect. 2.1.4, in contrast to Refs. [21, 35] where the quantities are micromagnetic. The degeneracy in the angle θ is lifted by the perpendicular field, so in this case $\theta = \pi/2 = \theta_A = \theta_S = \theta_B$. The potential is symmetric, with two minima of the energy at $\phi_A = \arcsin(h)$ and $\phi_B = \pi - \arcsin(h)$. The saddle point is found at $\phi_S = \pi/2$ (Fig. 3.6a). By applying Langer's theory, the attempt frequency for the magnetization reversal was derived in Ref. [35] and is expressed as

$$f_0 = \frac{\lambda_+}{2\pi}\Omega_0 = \frac{1}{2\pi}\left[\frac{\gamma_e K}{M_s(1+\alpha^2)}\left(\alpha(1-2h) + \sqrt{\alpha^2 - 4h(h-1)}\right)\right]\sqrt{1+\frac{1}{h}}. \tag{3.94}$$

In the previous expression, the gyromagnetic factor for the free electron, γ_e, and the damping factor, α, are as defined in Sect. 3.3.2. We used the following parameters: $N = 4 \times 4$, $a = 0.1$ nm, $\alpha = 0.5$, $\mathscr{K} = 1$ MJ/m^3, $\mathscr{M}_s = 0.5\mu_0^{-1}$ A/m, and $\gamma_e = 1.76 \times 10^{11}$ rad.T^{-1}.s^{-1}, where the anisotropy constant and saturation magnetization were chosen to match that of Ref. [21] and correspond to a small single-domain magnetic cube. We varied the reduced field h in $[0.01, 0.4]$ and computed an attempt frequency f_0 for each value of h. We obtained the graph in Fig. 3.6b, which shows good agreement with the analytical formula. This calculation was repeated with the

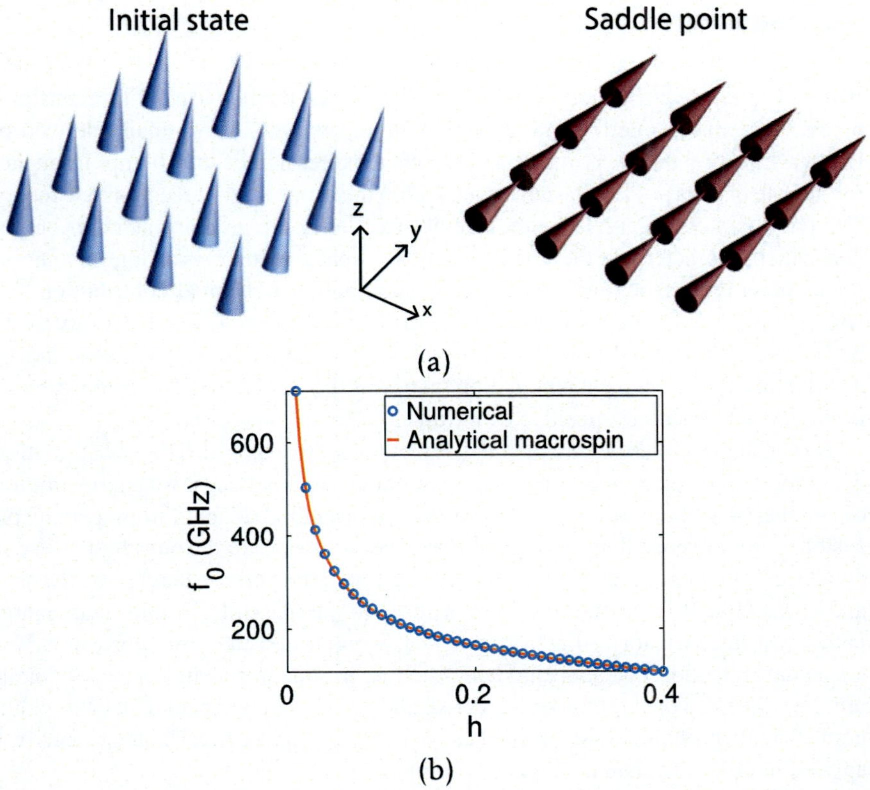

Fig. 3.6 **a** Initial stable state and saddle point of the magnetization reversal in a perpendicular field applied magnetic along y in a 4×4 spin system. **b** Attempt frequency of coherent magnetization reversal for different values of the reduced perpendicular applied field h. The analytical expression is derived for a single macrospin using Langer's theory

perpendicular field applied along z, in order to validate the treatment of the poles, and yielded the same result.

3.5.4.5 Skyrmion Collapse

We used the same parameters as given above for testing the skyrmion profile. The energy barrier for the isotropic collapse of the skyrmion was compared to one obtained with the implementations from Refs. [36] and [29]. The same barrier of $\Delta E = 2.83\ J_{\text{ex}}$ was found in all cases with good accuracy. The process of skyrmion collapse will be studied in more details in Chap. 4.

3.6 Summary

In this chapter, we presented the extension of Kramers' method to a multidimensional phase space in the intermediate-to-high damping regime, as originally derived by Langer. Langer's theory provides an analytical expression of the attempt frequency for the rate of decay of a metastable state within the framework of reaction rate theory. The two main contributions to the attempt frequency are an entropic contribution, Ω_0, which characterizes the change of phase space volume available to thermal fluctuations upon reaching the transition state, and a dynamical contribution, λ_+, which corresponds to the rate of growth of the (fastest growing) instability at the barrier top. We showed how the theory applies to magnetic spin systems and we derived an expression for the transition matrix of the LLG equation, from which λ_+ is extracted as its largest positive eigenvalue.

Next, the principles behind the GNEB method were exposed. The GNEB method is a numerical scheme to relax paths of minimum energy between two energy minima on an energy surface. It is a version of the NEB method adapted to magnetic spin systems, in the sense that the curved manifold is taken into account by the use of geodesics to calculate distances, and by the projection of the tangent to the path and of the GNEB force onto the tangent space to the manifold. A minimum energy path constitutes a path of highest statistical weight, and therefore corresponds to a specific transition mechanism. Using a climbing image scheme, it is possible to precisely identify the highest first-order saddle point along the path. Once the saddle point is known, the internal energy barrier is known, and Langer's theory can be a applied in order to calculate a rate prefactor.

Finally, we gave some details on the numerical implementation of GNEB and Langer's theory into an atomistic code. The rate calculation was successfully tested against the analytical expression of the rate of reversal of a single macrospin under a perpendicular magnetic field within the IHD regime. The methods presented in this chapter will be used in the rest of this thesis to compute annihilation rates of metastable magnetic skyrmions.

References

1. Coffey W, Kalmykov YP (2012) The Langevin equation: with applications to stochastic problems in physics, chemistry and electrical engineering, vol 27. World Scientific
2. Einstein A (1905) Über die von der molekularkinetischen Theorie der Wärme geforderte Bewegung von in ruhenden Flüssigkeiten suspendierten Teilchen [Investigations on the theory of Brownian Movement]. Annalen der Physik 17(8):549–560
3. Kubo R (1966) The fluctuation-dissipation theorem. Rep Prog Phys 29(1):255
4. Risken H (1996) The Fokker-Planck equation. Springer series in synergetic, vol 18. Springer, Berlin, Heidelberg
5. Langevin P (1908) Sur la Théorie de Mouvement Brownien. Comptes Rendus Académie des Sciences Paris 146:530–533
6. Hänggi P, Talkner P, Borkovec M (1990) Reaction-rate theory: fifty years after Kramers. Rev Modern Phys 62(2):251

7. Kramers HA (1940) Brownian motion in a field of force and the diffusion model of chemical reactions. Physica 7(4):284–304
8. Langer JS (1969) Statistical theory of the decay of metastable states. Ann Phys 54:258–275
9. Coffey W, Garanin D, McCarthy D (2001) Crossover formulas in the Kramers theory of thermally activated escape rates—application to spin systems. Adv Chem Phys 117:483–765
10. Becker R, Döring W (1935) Kinetische Behandlung der Keimbildung in übersättigten Dämpfen. Annalen der Physik 416(8):719–752
11. Bessarab PF, Uzdin VM, Jonsson H (2015) Method for finding mechanism and activation energy of magnetic transitions, applied to skyrmions and antivortex annihilation. Comput Phys Commun 196:335–347
12. Duff G (2008) Langer's method for the calculation of escape rates and its application to systems of ferromagnets. PhD thesis, Dublin Institute of Technology
13. Loxley P (2005) Nucleation of solitons in the presence of defects. PhD thesis, University of Western Australia
14. Braun H-B (1994) Kramers's rate theory, broken symmetries, and magnetization reversal. J Appl Phys 76(10):6310–6315
15. Bessarab PF, Müller GP, Lobanov IS, Rybakov FN, Kiselev NS, Jónsson H, Uzdin VM, Blügel S, Bergqvist L, Delin A (2018) Lifetime of racetrack skyrmions. Sci Rep 8(1):1–10
16. von Malottki S, Bessarab PF, Haldar S, Delin A, Heinze S (2019) Skyrmion lifetime in ultrathin films. Phys Rev B 99(6):060409
17. Bessarab PF, Uzdin VM, Jónsson H (2012) Harmonic transition-state theory of thermal spin transitions. Phys Rev B 85:184409
18. Private communication
19. García-Palacios JL, Lázaro FJ (1998) Langevin-dynamics study of the dynamical properties of small magnetic particles. Phys Rev B 58:14937–14958
20. Loxley PN, Stamps RL (2006) Theory for nucleation at an interface and magnetization reversal of a two-layer nanowire. Phys Rev B 73(2):024420
21. Fiedler G, Fidler J, Lee J, Schrefl T, Stamps RL, Braun H, Suess D (2012) Direct calculation of the attempt frequency of magnetic structures using the finite element method. J Appl Phys 111(9):093917
22. Mills G, Jónsson H, Schenter GK (1995) Reversible work transition state theory: application to dissociative adsorption of hydrogen. Surf Sci 324(2–3):305–337
23. Jónsson H, Mills G, Jacobsen KW (1998) Nudged elastic band method for finding minimum energy paths of transitions. In: Classical and quantum dynamics in condensed phase simulations, pp 385–404
24. Henkelman G, Uberuaga BP, Jónsson H (2000) A climbing image nudged elastic band method for finding saddle points and minimum energy paths. J Chem Phys 113(22):9901–9904
25. Cortés-Ortuño D, Wang W, Beg M, Pepper RA, Bisotti M-A, Carey R, Vousden M, Kluyver T, Hovorka O, Fangohr H (2017) Thermal stability and topological protection of skyrmions in nanotracks. Sci Rep 7(1):4060
26. Henkelman G, Jónsson H (2000) Improved tangent estimate in the nudged elastic band method for finding minimum energy paths and saddle points. J Chem Phys 113(22):9978–9985
27. Aharoni A (2000) Introduction to the theory of ferromagnetism. International series of monographs on physics, vol 109. Clarendon Press, Oxford
28. Swope WC, Andersen HC, Berens PH, Wilson KR (1982) A computer simulation method for the calculation of equilibrium constants for the formation of physical clusters of molecules: application to small water clusters. J Chem Phys 76(1):637–649
29. Müller GP, Hoffmann M, Dißelkamp C, Schürhoff D, Mavros S, Sallermann M, Kiselev NS, Jónsson H, Blügel S (2019) Spirit: multifunctional framework for atomistic spin simulations. Phys Rev B 99(22):224414
30. Galassi M, Davies J, Theiler J, Gough B, Jungman G, Alken P, Booth M, Rossi F, Ulerich R (2002) GNU scientific library. Network Theory Limited
31. Qiu Y, Guennebaud G, Niesen J (2015) Spectra: C++ library for large scale eigenvalue problems
32. Guennebaud G, Jacob B et al (2010) Eigen v3. http://eigen.tuxfamily.org

33. Vansteenkiste A, Leliaert J, Dvornik M, Helsen M, Garcia-Sanchez F, Van Waeyenberge B (2014) The design and verification of MuMax3. AIP Adv 4(10):107133
34. Heo C, Kiselev NS, Nandy AK, Blügel S, Rasing T (2016) Switching of chiral magnetic skyrmions by picosecond magnetic field pulses via transient topological states. Sci Rep 6:27146
35. Schratzberger J, Lee J, Fuger M, Fidler J, Fiedler G, Schrefl T, Suess D (2010) Validation of the transition state theory with Langevin-dynamics simulations. J Appl Phys 108(3):033915
36. Heistracher P, Abert C, Bruckner F, Vogler C, Suess D (2018) GPU-accelerated atomistic energy barrier calculations of skyrmion annihilations. IEEE Trans Magn 54(11):1–5

Chapter 4
Thermal Stability of Chiral Magnetic Skyrmions from Langer's Theory

4.1 Introduction

In what follows, we apply Langer's theory adapted to magnetic spin systems to compute annihilation rates of chiral metastable magnetic skyrmions within the atomistic framework. The computation is carried out in two main steps. Firstly, the GNEB method is used to compute a minimum energy path through the energy landscape, and the saddle point is identified with high accuracy. Secondly, a rate prefactor is calculated according to Eq. (3.31). The results presented in this chapter are based on original publications [1, 2].

The chapter is organized as follows. In Sect. 4.2, we present the system under study and give the material parameters. In Sect. 4.3, we show paths to annihilation and corresponding energy profiles of isolated skyrmions via collapse and escape processes. After that, we compute transition rates associated with each mechanism in Sect. 4.4, and we consider the role of local eigenmodes and the entropic contribution to the prefactor. Finally, the results are discussed in Sect. 4.5.

4.2 Simulated System

We simulate a two-dimensional system representing a thin magnetic film. The atomistic Hamiltonian is as defined in Sect. 2.1. The total simulated domain contains 50×50 spins, or 80×80 spins in the configuration involving two skyrmions, and we keep open (non-periodic) boundary conditions. The exchange interactions are limited to first nearest-neighbours. As we explained in Sect. 2.1.1, Néel-type skyrmions are stabilized by interfacial DMI, and Bloch-type skyrmions are stabilized by bulk-type DMI. As for antiskyrmions, they are favoured by modifying the interfacial DMI such that $\mathbf{D}_{ij} = -D\mathbf{r}_{ij} \times \mathbf{e}_z$ when $\mathbf{r}_{ij} = \mathbf{e}_y$ [3]. We use an isotropic exchange constant of $J_{\text{ex}} = 1.6 \times 10^{-20}$ J ($\sim$100 meV), which corresponds to an exchange stiffness of

L. Desplat, *Thermal Stability of Metastable Magnetic Skyrmions*, Springer Theses,
https://doi.org/10.1007/978-3-030-66026-0_4

$A = 16$ pJ/m with lattice constant $a = 1$ nm, and the saturation magnetization is set to $\mathcal{M}_s = 1.1 \times 10^6$ A m^{-1}. These values correspond to typical Pt/Co(0.6 nm)/AlO$_x$ samples, in which interfacial DMI arises from the symmetry breaking at the interface [4]. The reduced parameters are then adjusted in order to stabilize small isolated metastable skyrmions at zero-field, with a diameter of approximately ten lattice sites, and are the following: $(b, d, k) = (0, 0.36, 0.4)$ [5]. In what follows, we keep the external magnetic field to zero unless stated otherwise. The damping term in the LLG equation (Eq. (3.46)) is set to $\alpha = 0.5$, which corresponds to commonly found values for ultrathin magnetic films with DMI [4], while pertaining to the IHD regime of Langer's theory. The gyromagnetic ratio is that of the free electron, $\gamma_e = 1.76 \times 10^{11}$ rad s^{-1} T^{-1}. The GNEB scheme is used on $Q = 10$ or 15 images of the system for single-skyrmion mechanisms, and $Q = 6$ images for the two-skyrmion mechanism.

4.3 Paths to Annihilation

4.3.1 Collapse Mechanisms

4.3.1.1 Isolated Skyrmion

We first study the case of the collapse of a single isolated skyrmion. At zero-field, the core of the skyrmion coincides with a lattice site (image (im.) 1 in Fig. 4.1a). In that context, we report two separate mechanisms for the collapse of the Néel skyrmion. We show the spin profiles at selected images in Fig. 4.1a and b, and the corresponding energy profiles in Fig. 4.2a and b.

In the first case, which we refer to as mechanism n°1, the skyrmion progressively shrinks in size without breaking the cubic symmetry, while its core remains centred on a lattice site (Fig. 4.1a). This is accompanied by a slow increase in energy (Fig. 4.2a). At the saddle point (im. 8 in Fig. 4.1a, and inset in the middle in Fig. 4.2a), which we refer to as S_1, the core spin still points along $-z$, while its first neighbours are aligned in-plane. Once the core begins to reverse, the cubic symmetry is broken: the remaining spins flip, the energy drops dramatically and the system reaches the ferromagnetic ground state. The corresponding energy barrier is $\Delta E_{\mathrm{col},1} = 2.83\ J_{\mathrm{ex}}$. This corresponds to about $10 k_B T_{300}$ with the present value of J_{ex}, where the notation T_{300} is used to indicate $T = 300$ K.

Alternatively, if mechanism n°2 is realized, the skyrmion core may shift to an interstitial position in the centre of a square unit cell and, from there, the skyrmion uniformly shrinks in size (Fig. 4.1b). The saddle point configuration, which we call S_2, corresponds to the four central spins pointing towards the centre of the unit cell (im. 8 in Fig. 4.1b, and inset in Fig. 4.2b). The energy profile is similar to that of the first mechanism. This particular path to collapse was first reported in Ref. [6]. In the present case, however, the translational modes at the SP become unstable, resulting in a third-order saddle point. For that reason, while the GNEB scheme is able to relax

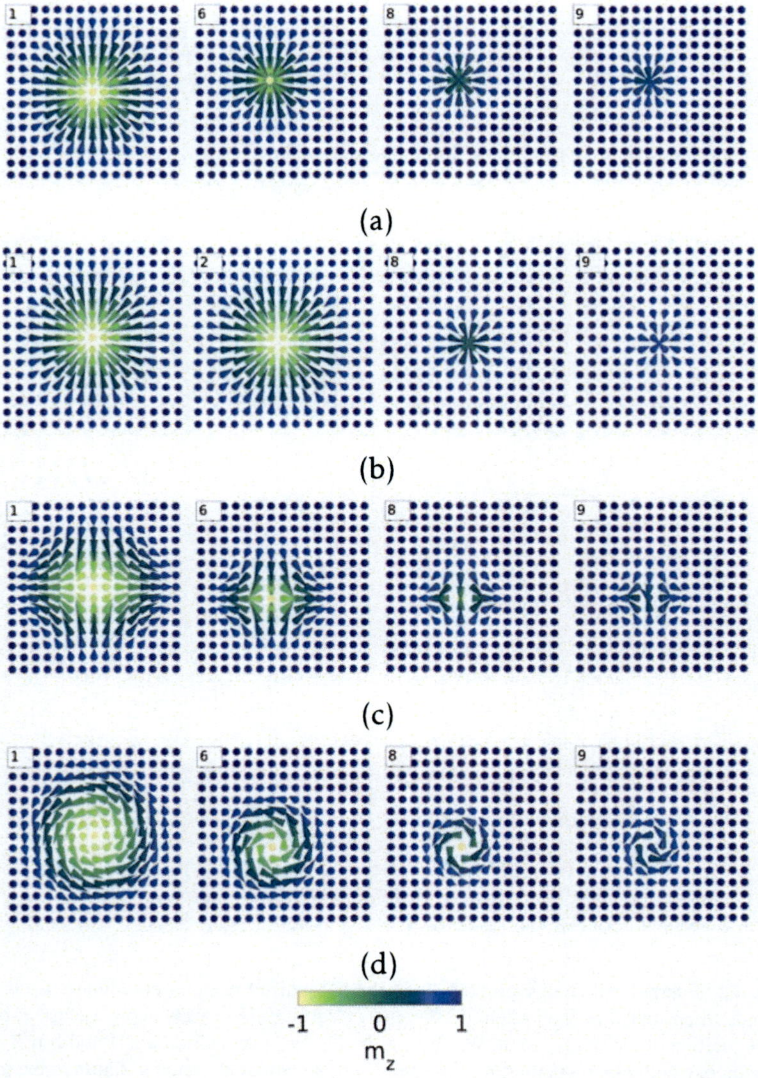

Fig. 4.1 Spin maps of the isotropic collapse mechanisms. The index in the top left-hand corner corresponds to the image index of the GNEB method. We show **a**, **b** for a Néel skyrmion, **a** collapse mechanism n°1, and **b** collapse mechanism n°2, followed by **c**, **d** mechanism n°1 for **c** an antiskyrmion and, **d** a Bloch skyrmion

the path down to a low tolerance, the CI-GNEB scheme tends to shift it back to the first mechanism. We still manage to relax the path down to a tolerance on the GNEB force of $F_{\text{tol}} = 9 \times 10^{-7}$. The influence of the set tolerance on the energy barrier and the attempt frequency is briefly discussed in Appendix A. The corresponding energy barrier is similar to that of the first mechanism: $\Delta E_{\text{col},2} = 2.90\ J_{\text{ex}}$. Since the

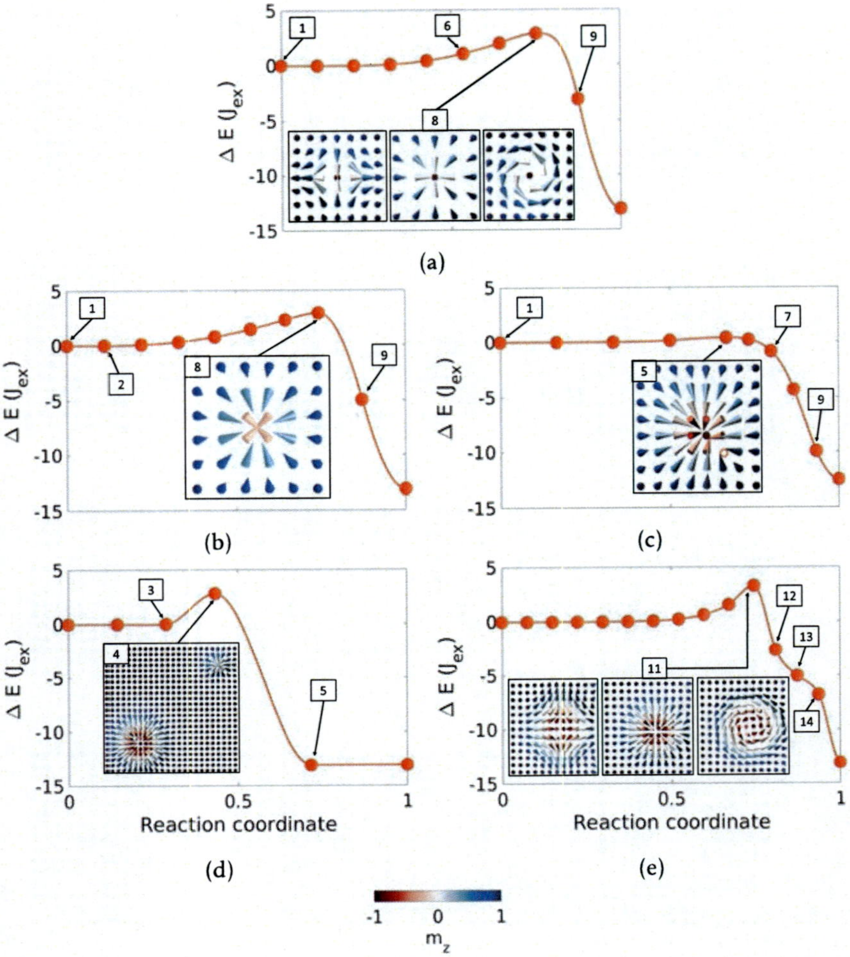

Fig. 4.2 Interpolated energy profiles along the normalized reaction coordinate. Each dot corresponds to an image of the system on the energy surface. The insets show a closeup of the spin configuration at the saddle point. We show **a**, **b** collapse mechanisms **a** n°1 and **b** n°2 of a single skyrmion, **c** collapse in the presence of a single non-magnetic defect, **d** collapse in the presence of another skyrmion and, **e** escape through a flat boundary. In **a** and **e**, the inset respectively shows the saddle point configurations of the antiskyrmion, the Néel skyrmion, and the Bloch skyrmion

translational modes at the saddle point are not Goldstone modes [1, 7], S_1 and S_2 should indeed be treated as separate states, and they are therefore associated with different activation rates. S_2 is found above S_1 on the energy surface, by advancing along the eigenbasis coordinate associated with a translation mode.

Lastly, the collapse of an antiskyrmion and of a Bloch skyrmion, following a similar path to that of mechanism n°1, are shown in Fig. 4.1c and d, respectively. They both exhibit a very similar behaviour to that of the Néel skyrmion, with a breaking of the symmetry past the saddle point corresponding to the flipping of the spin at the core. The energy profile along the path and the activation energy are the same for all three types of skyrmions. We did not calculate the collapse of the Bloch skyrmion and of the antiskyrmion via mechanism n°2, but we believe it would also be similar to that of the Néel skyrmion.

4.3.1.2 Effect of a Non-magnetic Defect

The effect of a single non-magnetic defect on the collapse was studied. This is important, because real systems are not perfect, and it has been experimentally observed that atomic defects are preferred sites for nucleation, pinning, and annihilation of skyrmions [8, 9]. In Pd/Fe/Ir(111), the interaction of skyrmions with atomic-scale defects was studied via spin-polarized scanning tunneling microscopy (SP-STM) [9]. Native inlayer defects were found in the Pd layer and were assumed to be single Fe atoms due to intermixing. The defects were reported to be slightly off-centre compared to the pinned skyrmion. This was shown to be the case in bigger skyrmions either due to reduced spin stiffness or reduced DMI at the defect position, or due to reduced anisotropy for any skyrmion size. Here, we chose to follow Ref. [10] by simulating a non-magnetic defect. This is done by setting the anisotropy to zero at a given site, as well as all exchange interactions with the neighbouring sites. Im. 1 in Fig. 4.3a shows that the skyrmion relaxes such that the defect sits where the spins lie in-plane, which reproduces experimental observations [9]. The simulated skyrmion is small, so the off-centre pinning is likely due to a reduction in the anisotropy energy. We show spin maps at selected images in Fig. 4.3a, and the corresponding energy profile in Fig. 4.2c. As the core moves towards the defect—which takes place along the diagonal of the square lattice—the skyrmion shrinks in size. This costs little energy compared to the isotropic collapse mechanism, as seen on the energy profile of Fig. 4.2c. The saddle point is im. 5 along the path and corresponds to a larger skyrmion than in the defect-free configuration, but the skyrmion is deformed and rendered asymmetric by the defect. Consistent with results presented in Ref. [10], the presence of the defect significantly lowers both annihilation and nucleation barriers, and we obtain $\Delta E_{\mathrm{def}} = 0.38\ J_{\mathrm{ex}}$, which is one order of magnitude lower than $\Delta E_{\mathrm{col},1,2}$.

4.3.1.3 Effect of a Second Skyrmion

We consider what happens when two skyrmions get close to each other. We show spin maps at selected images in Fig. 4.3b, and the corresponding energy profile in Fig. 4.2d. When the skyrmion cores are initially aligned along the lattice directions x or y, we observe that they rotate in order to approach each other along the lattice diagonal. Presumably, this is more energetically favourable due to the choice of first-neighbour interactions on the simple square lattice. Consequently, we simply initialize the skyrmions diagonally from each other. Note that this may be less true

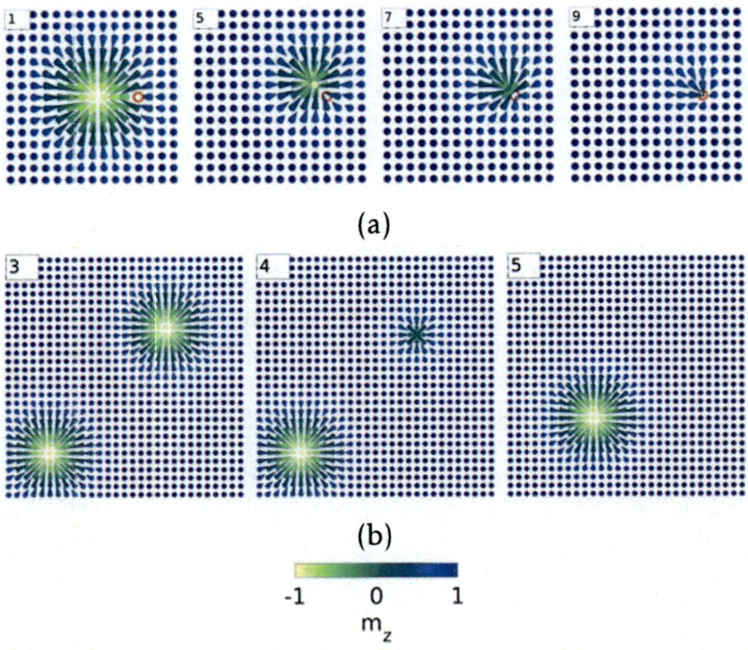

Fig. 4.3 Spin maps of the collapse mechanisms of a Néel skyrmion in the presence of **a** a non-magnetic defect, where the position of the defect is indicated by a red circle, and **f** a second skyrmion. The conventions are the same as in Fig. 4.1

in real systems with corrections in the exchange to include interactions beyond the first neighbours, and in systems where the skyrmions are big compared to the lattice parameter and decouple from the lattice. Additionally, interface skyrmions tend to be observed in systems with a hexagonal lattice, for instance Pd/Fe and Rh/Co on Ir(111) [8, 11], in which these considerations do not apply. We set the transition path for a merging of the two skyrmions into one, as observed experimentally in the case of the decay of a skyrmion lattice into the helical state [12]. However, the search for a first-order SP consistently results in a switch in mechanism, and the collapse of one of the skyrmions is relaxed instead of the merging (Fig. 4.3b). This behaviour hints at the fact that the merging mechanism involves a higher-order SP, which may therefore involve a higher barrier and be less favourable than the collapse. It was confirmed by a mode following method [13] that, with our present set of parameters, the merging mechanism is indeed associated with a higher order SP. However, for other parameters, the path to merging may in fact traverse a first-order SP [13].

The mechanism is as follows. The skyrmions initially approach one another, which at first costs almost no energy (im. [1–3] in Fig. 4.2d), until a critical distance is reached at which the collapse of the upper skyrmion is initiated. The saddle point is the same as that of collapse mechanism n°1 for the upper skyrmion, while the other one remains stable (im. 4 in Fig. 4.3b, and inset in Fig. 4.2d). At finite temperature, isolated skyrmions perform a diffusive Brownian motion on the lattice under the effect of thermal fluctuations [14]. Although the interaction potential between chiral skyrmions is repulsive at all distances [15], some fluctuations may bring them close to each other. In that case, it is interesting to study how the collapse rate is affected. The activation energy for collapse remains the same: $\Delta E_{2sk} = 2.82\ J_{ex}$. We shall see how the attempt frequency is modified in Sect. 4.4.

4.3.2 Escape Through a Boundary

4.3.2.1 Flat Boundary

Escape through a boundary is another possible path in finite-size systems. We show spin maps at selected images in Fig. 4.4a, b and c respectively for a Néel skyrmion, an antiskyrmion and a Bloch skyrmion, and the corresponding energy profile is shown in Fig. 4.2e. Because of the DMI, the spins are tilted at the boundary with the same chirality as that of the skyrmion. For the same reason that chiral skyrmions repel each other, the boundary is also repulsive to the skyrmion [16, 17], such that the total energy increases as the skyrmion leaves the centre of the lattice and moves towards an edge (ims. [1–11] in Fig. 4.2e). For the Néel skyrmion, the antiskyrmion, and the Bloch skyrmion, the saddle point corresponds to a position where the skyrmion sits adjacent to the boundary (im. 11 in Fig. 4.4a, b and c, respectively, and insets in Fig. 4.2e), as was also shown in Ref. [10]. Past the saddle point, the skyrmion deforms and elongates as it comes in contact with the edge and begins to disappear. This is accompanied by a large decrease in the energy. Im. 13 in Fig. 4.4a, b, c and 4.2e corresponds to a half-skyrmion—or meron—sitting on the edge. In the vicinity of this point, the decrease in energy appears to slow down, before speeding up again as the rest of the remaining skyrmion disappears. The activation energy obtained for this mechanism is the highest one of the three processes studied here, although they are all of a similar magnitude of 2-3 J_{ex}: $\Delta E_{esc} = 3.28\ J_{ex}$, once again for all three types of skyrmions. While it should be noted that in the present configuration, this mechanism possesses four equivalent realisations (one at each side of the square), we shall not include a factor of four in the attempt frequency, as the escape rate through one edge is potentially more relevant that the total escape rate out of a square sample.

4.3.2.2 Curved Boundary

Figure 4.4d shows the escape of a Néel skyrmion through a curved boundary. For that purpose, a disc-shaped system is obtained from the square system by converting outer spins into non-magnetic sites. The nature of the discrete lattice means that the boundary exhibits a staircase effect, and so the spin lattice as a whole only exhibits a cubic-type of symmetry, as opposed to a radial symmetry. Note that the boundary is considered curved if the curvature is perceived at the scale of the skyrmion, which is

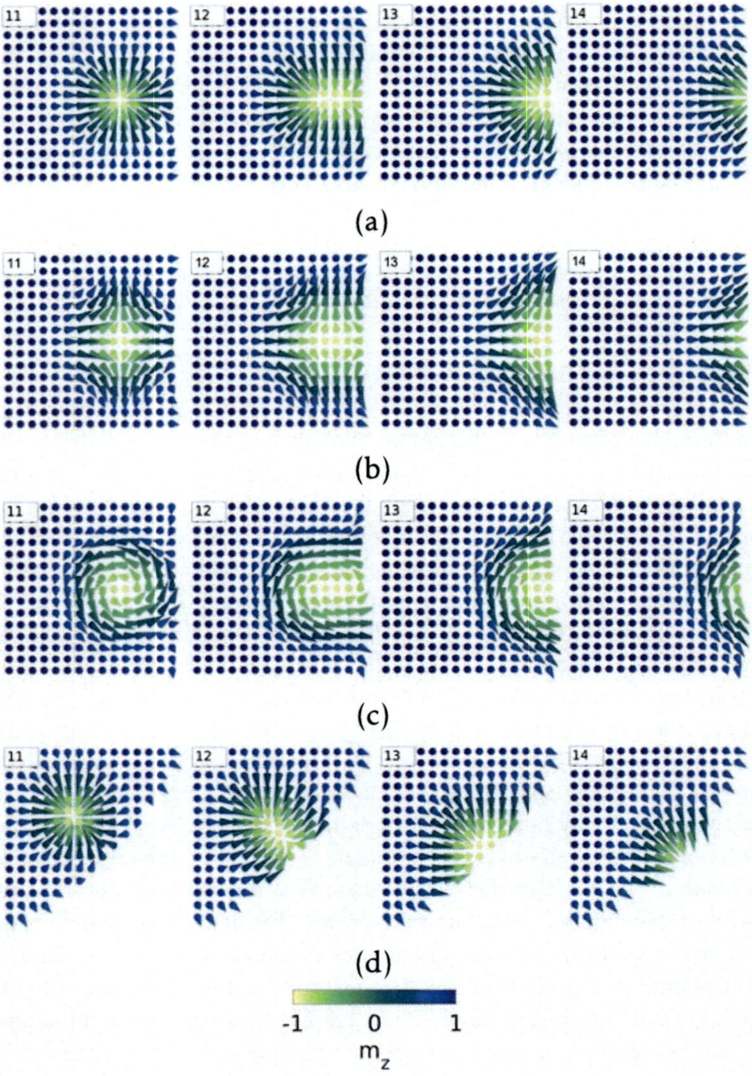

Fig. 4.4 Spin maps of the escape mechanisms. The conventions are the same as in Fig. 4.1. On all the subfigures, the saddle point is im. 11, which is the state where the skyrmion sits tangent to the boundary. We show **a** a Néel skyrmion, **b** an antiskyrmion, **c** a Bloch skyrmion, and **d** a Néel skyrmion escaping through a curved boundary

the case here. The escape process is similar to that of the flat boundary configuration, but the activation energy is found to be slightly higher: $\Delta E_{\mathrm{curv}} = 3.60\ J_{\mathrm{ex}}$. The energy profile is similar to that of Fig. 4.2e, and so we do not show it.

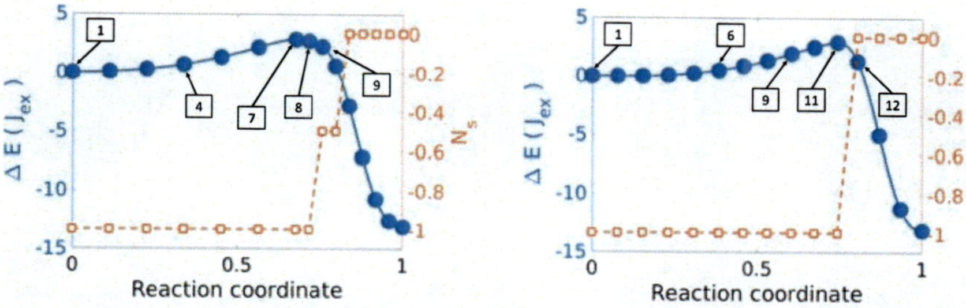

(a) (b)

Fig. 4.5 Interpolated energy profile and evolution of the topological charge N_s along the reaction coordinate for collapse mechanisms **a** n°1 and **b** n°2. N_s is calculated according to Eq. (2.38)

4.3.3 Paths to Collapse and Evolution of the Topological Charge

We compare the evolution of the topological charge N_s along the path for collapse mechanisms n°1 and 2. To do so, we compute a value of N_s at each image using the definition from Eqs. (2.38) and (2.39). We plot the result in Fig. 4.5 as well as the energy profiles. We now use $Q = 15$ images to improve the definition of the path in the area where the topological charge drops. In Fig. 4.6, we use the "ball in a net" type of representation from Sect. 2.3.2 to show the spin configuration at different images along the MEP. As previously described in Sect. 2.3.2, vertices correspond to the tip of the magnetic vectors with their origin in the centre of the sphere, and the edges represent the exchange coupling between first neighbours. The ferromagnetic background (spins along $+z$) is found at the north pole of the sphere, while the core of a skyrmion typically points along the south pole (along $-z$). The view is set just below the south pole, looking axially towards the $+z$ direction, such that the core of the skyrmion is at the centre of the view. Since we defined the topological charge on the discrete lattice as the number of times the magnetization "net" wraps around the unit sphere, this type of representation provides an intuitive way to visualise the evolution of the topological charge.

For mechanism n°1, we find that the topological charge is conserved past the saddle point, which is im. 7 (Figs. 4.5a and 4.6a), and until im. 8. At ims. 9 and 10, the core spin starts to flip, and we find a half-integer topological charge: $N_s = -0.5$. After that, the topological charge vanishes and the configuration becomes topologically trivial.

On the other hand, during the second mechanism, the topological charge drops from $N_s = -1$ to 0 at im. 12, which is the image directly following the one at the saddle point (im. 11 in Figs. 4.5b and 4.6b). However, a half-integer value may exist as a punctual state for this mechanism (punctual in the sense that this state would have zero-length along the reaction coordinate) and would correspond to the case where the four spins in the centre point exactly in-plane. Therefore, the two collapse

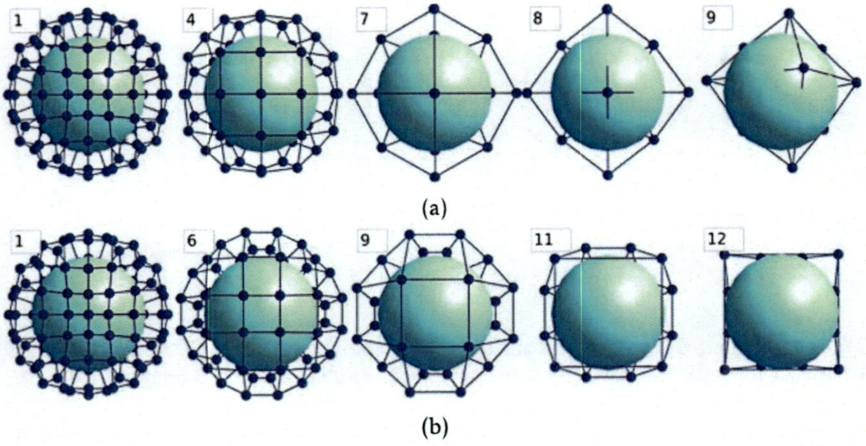

Fig. 4.6 "Ball in a net" representation of the spin configuration at selected images along the path for collapse mechanisms **a** n°1 and **b** n°2. The image index is given in the top left-hand corner. The topological charge on the discrete spin lattice is defined as the number of times the magnetization net wraps around the unit sphere, so this representation shows the evolution of the topological charge

mechanisms provide an example of two different ways through which an integer topological charge may be annihilated without escaping through the system's edges.

4.4 Transition Rates from Langer's Theory

4.4.1 Overview

Once the saddle point is accurately identified along a path, the corresponding rate prefactor can be calculated according to Langer's theory (Eq. (3.31)). All terms entering in the calculation of activation rates are summarized in Table 4.1 for each mechanism. In Fig. 4.7, we show the evolution of the average lifetime of the skyrmion, $\tau = k^{-1}$, associated with each annihilation mechanism as a function of temperature. We find that the collapse on a non-magnetic defect yields by far the shortest lifetime. At low temperature, collapse mechanisms are found more probable than escape, but this trend reverses at higher and room temperature, where the longest lifetimes are associated with collapse. In the absence of defects, the average waiting times between two similar annihilation events at room temperature (300 K) are found between 0.3 ms, which is associated with the escape through a flat boundary, and 21 ms, which corresponds to collapse mechanisms n°2. Collapse mechanism n°1 and escape through a flat boundary were studied for the Néel and Bloch skyrmions, as well as the antiskyrmion, and yield the same results. In the cases of escape through a curved boundary, collapse mechanism n°2, and collapse on a defect, we studied only

Table 4.1 Terms of the rate prefactor and total annihilation rate at T=300 K for (a) collapse processes and (b) escape processes. From left to right, we give the internal energy barrier, ΔE, in units of the isotropic exchange constant, J_{ex}, the entropic contribution to the prefactor, $\Omega_0 = \Omega_{0,\mathrm{tot}}$, the dynamical contribution to the prefactor, λ_+, the Arrhenius prefactor, f_0, and the total rate of annihilation at room temperature, $k_{300} = k\,(\mathrm{T} = 300\,\mathrm{K})$. $\Omega_{0,\mathrm{int}}$ gives the contribution of internal modes to the prefactor and $\Omega_{0,\mathrm{tot}}$ gives the total contribution of all modes. Their ratio shows that the main contribution comes from the internal modes. f_0 and k_{300} are calculated using $\Omega_0 = \Omega_{0,\mathrm{tot}}$. The entropic contribution, Ω_0, is found to be much smaller than one, which is a stabilizing contribution. This is the reason why the obtained attempt frequencies are typically found in the MHz regime, which is uncharacteristically low for magnetic structures

(a) Collapse

Mechanism	ΔE (J_{ex})	$\Omega_{0,\mathrm{int}}$	$\Omega_{0,\mathrm{tot}}$ ($\times 10^{-5}$)	$\frac{\Omega_{0,\mathrm{int}}}{\Omega_{0,\mathrm{tot}}}$	λ_+ (GHz)	f_0 (MHz)	k_{300} (kHz)
single sk. 1	2.83	0.0015	3.51	43	1200.47	6.70	0.12
single sk. 2	2.90	0.00048	1.05	46	2124.23	3.54	0.05
defect	0.38	0.0214	116	18	145.41	26.90	6190.40
two sk.	2.82	0.0009	2.32	42	1200.23	4.43	0.08

(b) Escape

Mechanism	ΔE (J_{ex})	$\Omega_{0,\mathrm{int}}$	$\Omega_{0,\mathrm{tot}}$ ($\times 10^{-3}$)	$\frac{\Omega_{0,\mathrm{int}}}{\Omega_{0,\mathrm{tot}}}$	λ_+ (GHz)	f_0 (MHz)	k_{300} (kHz)
flat bound.	3.28	0.0349	12.4	2.8	522.94	1036.2	3.25
curved bound.	3.60	0.0198	4.48	4.4	501.45	357.2	0.32

the Néel skyrmion, but it seems reasonable to assume the following results also hold for other types of skyrmion solutions. A possible issue of non-negligeable numerical rounding errors affecting the accuracy of the ratio of eigenvalues was previously mentioned in [18]. In Table A.1 of Appendix A, we gather results of simulations performed for different lattice sizes. We find that, as long as the skyrmion is not constrained by the boundary, the size of the lattice does not significantly affect the computed attempt frequencies, and therefore there seems to be no significant error originating from numerical noise in our results.

4.4.2 *The Thermal Role of Internal Eigenmodes*

4.4.2.1 Mode Classification

For each mechanism considered here, the curvatures of the energy surface at A and S are ordered by increasing amplitude with corresponding index i, and plotted in Fig. 4.8 for all i and in Fig. 4.9a–c, j–l for the first 15 or 25 values of i. Figure 4.9d–f, m–o show the ratio of individual curvatures, $\lambda_i^A/|\lambda_i^S|$, plotted in semilog scale. The value of

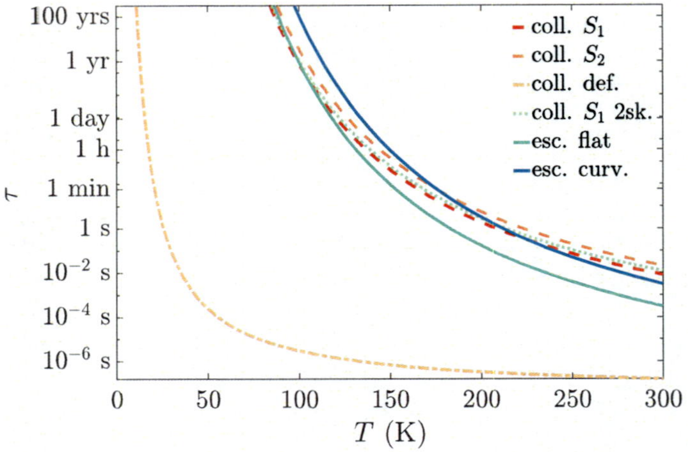

Fig. 4.7 Average lifetime of a skyrmion as a function of temperature with respect to all considered mechanisms calculated via Langer's theory as given in Eq. (3.31). The energy barriers and the detail of all the different contributions to the prefactor are given in Table 4.1. Here, the collapse on a non-magnetic defect yields by far the shortest lifetime. At low temperature, collapse mechanisms are found more probable than escape, but this trend reverses at room temperature, where the longest lifetimes are associated with skyrmion collapse. In the absence of defects, the average lifetimes at room temperature (300 K) are found between 0.3 ms (escape through a flat boundary) and 21 ms (collapse mechanisms n°2)

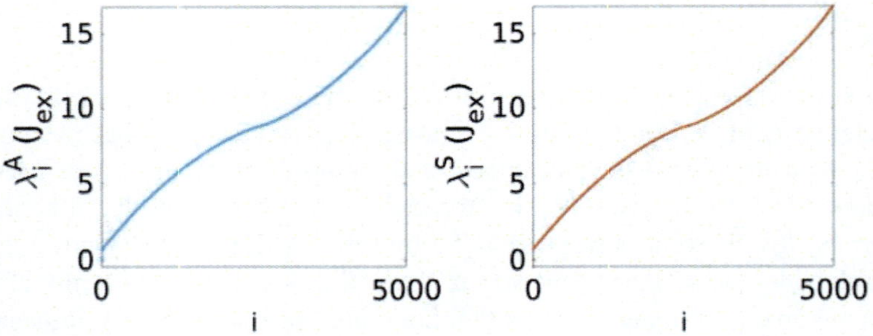

Fig. 4.8 All energy curvatures at A (in blue) and S (in red) ordered by increasing amplitudes for the case of the collapse of a single skyrmion via mechanism n°1. The other mechanisms exhibit the same profiles which we interpret as the dispersion of spin-wave excitations, with the exception of the first few eigenvalues. These are found below the main curve and are shown in Fig. 4.9a–c, j–l

$$\Omega_{0,i} = \sqrt{\prod_{j=1}^{i} \frac{\lambda_j^A}{|\lambda_j^S|}} \tag{4.1}$$

is plotted in a similar fashion in Fig. 4.9g–i, p–r, where all the curvatures are shown in the inset. We note that for all mechanisms, the ratio of curvatures only shows significant variations for small i and weak variations for $i \gg 1$. Consequently, the

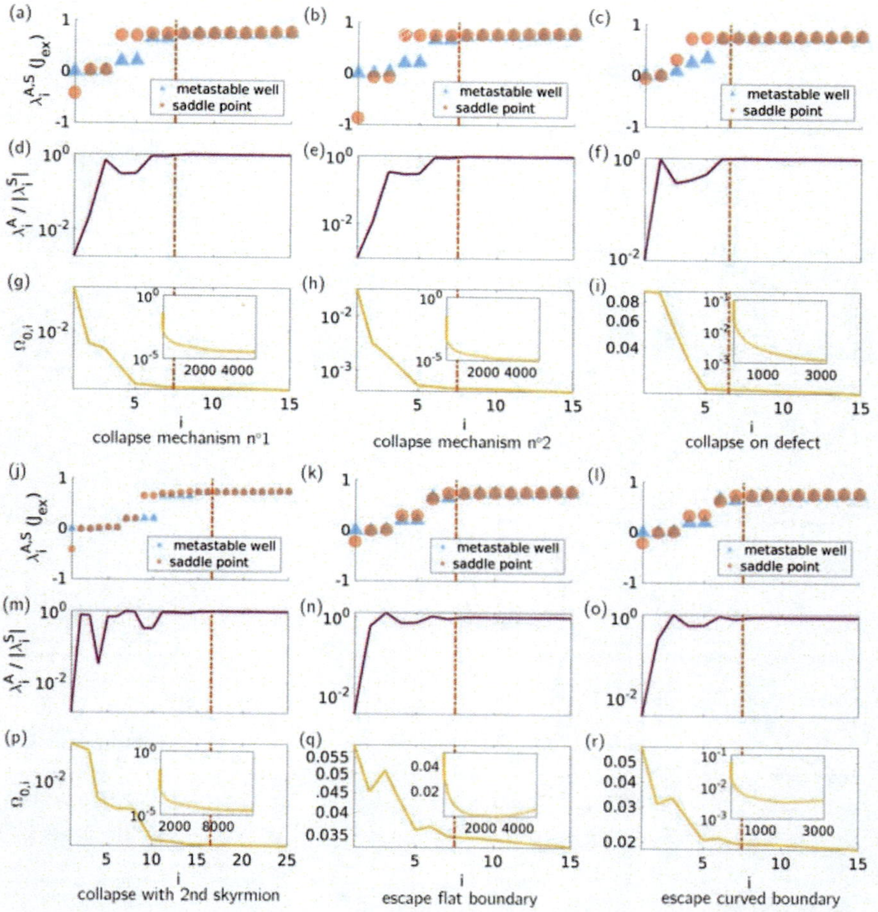

Fig. 4.9 Detail of the computation of the entropic contribution to the prefactor, Ω_0, for different annihilation mechanisms. The mechanism in question in indicated at the bottom of each set of three graphs. **a–c, j–l** We show the first 15 or 25 eigenvalues of the Hessian matrix of the energy evaluated at A (in blue) and S (in red). The eigenvalues are in unit of the isotropic exchange constant J_{ex} and ordered by increasing amplitudes, where the index i is attributed accordingly. A negative eigenvalue corresponds to an unstable mode of fluctuations. The red dotted line marks the separation between localized and collective eigenmodes. **d–f, m–o** Ratio of individual eigenvalues in semilog scales. The ratio of eigenvalues associated with local modes (left-hand side of the red line) is typically different from one, while for collective modes (right-hand side of the red line) it is close to one. **g–i, p–r** Cumulative $\Omega_{0,i}$ as defined in Eq. (4.1) in semilog scale, where the inset shows its value for all $i = 1 \dots 2N$. The value of Ω_0 used in the prefactor is $\Omega_{0,2N}$. The local modes contribute the most to Ω_0, and $\Omega_0 \ll 1$, which is a stabilizing contribution to the attempt frequency

value of $\Omega_{0,i}$ shows a strong i-dependence at small i and a weak dependence for larger i. More specifically, for all collapse mechanisms, $\Omega_{0,i}$ shows a strong i-dependence for small i, a medium dependence for intermediate i and a weak dependence for

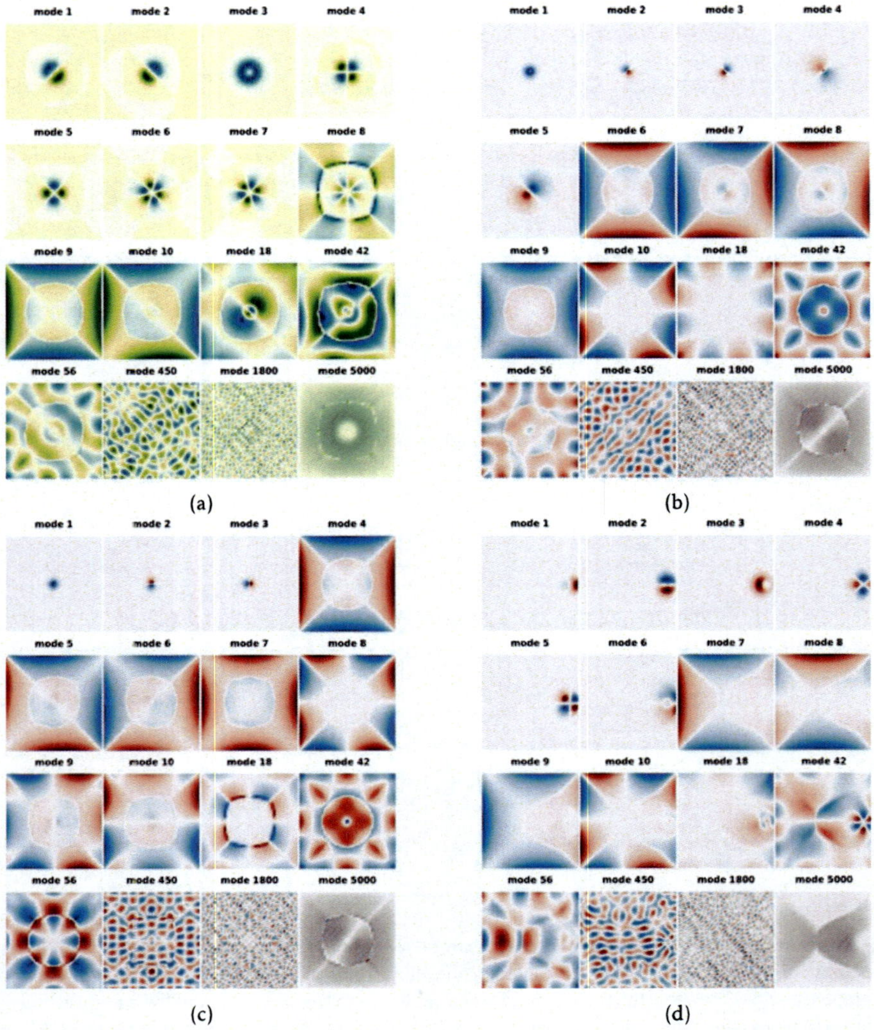

Fig. 4.10 Eigenmodes associated to the θ variable for a single skyrmion in a simulated system of 50×50 spins. The blue and green colour scheme is associated with metastable states and the blue and red one with saddle points. Negative amplitudes are plotted in blue and positive ones in green/red. The range of the colour map is adjusted on each plot so that zero-amplitude fluctuations coincide with white. Modes are designated via the i-index of the corresponding ordered eigenfrequencies of Fig. 4.9. We designate by i_{col} (short for "collective") the index of the first non-local mode. We show **a** the metastable single-skyrmion state with $i_{col} = 8$, **b** the saddle point of collapse mechanism n°1 with $i_{col} = 6$, **c** the saddle point of collapse mechanism n°2 with $i_{col} = 4$, and **d** the saddle point of the escape mechanism with $i_{col} = 7$

large i. In the case of escape mechanisms, the i-dependence decreases at low and intermediate indices, and goes up again in the domain of highest curvatures.

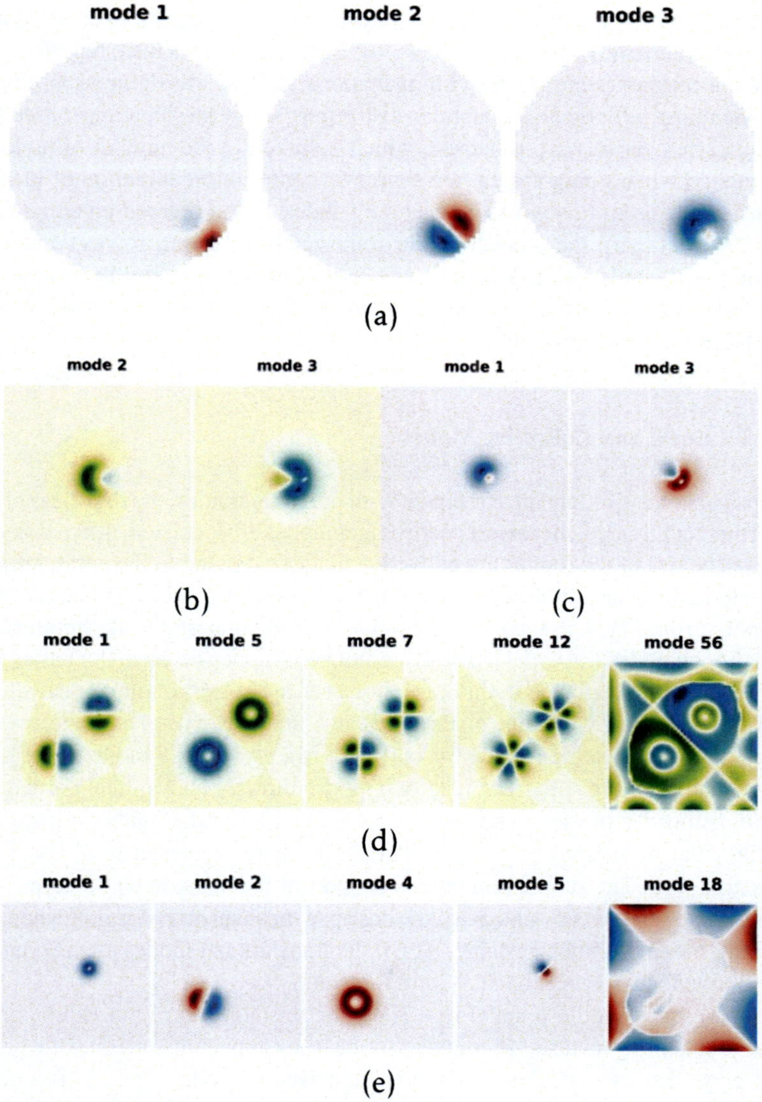

Fig. 4.11 Eigenmodes associated to the θ variable. The colour code is the same as in Fig. 4.10. We show **a** some modified internal modes for the escape through a curved boundary at S, **b**, **c** some modified internal modes for the collapse on a defect at A and S, and **d**, **e** some internal and collective modes of two coupled skyrmions on a 50×50 lattice at A and S

From Eq. (3.13), the relative amplitudes of small fluctuations about the extrema for each mode i are given by the components of the corresponding eigenvector, χ_i. This allows us to plot the spatial profiles for the θ-eigenmodes at A and S in Figs. 4.10 and 4.11. The ϕ-profiles exhibit similar behaviour and do not provide any further information for the following analysis. In recent years, the eigenmodes of skyrmions have been extensively investigated [19–29]. Localized modes were reported to exist below the magnon continuum, with excitations such as translational motion, uniform breathing, elliptic and triangular distortions, etc. Classifications based on the azimuthal number m were proposed, which is linked to the number of nodes $2|m|$ encountered when going around the skyrmion centre in the direction of increasing azimuth angle φ. In this work, we use the i-index of the ordered eigenfrequencies (Fig. 4.9) to classify the modes. While some relation to spin waves calculated in previous work could be made, our classification scheme is useful because it allows comparison of eigenvalues and eigenfunctions of the fluctuations around both the metastable state and the saddle point.

4.4.2.2 Local and Collective Modes

In Figs. 4.10 and 4.11, and in agreement with previous studies, the first clear observation is that for all cases, the lowest energy eigenmodes are localized, internal skyrmion modes. The rest of the modes are collective modes extended to the entire lattice and can be thought of as amplitudes corresponding to spin-wave (SW) excitations.

Firstly, from Fig. 4.10a, the first two modes of the metastable skyrmion state are translational modes ($m = \pm 1$). These are not zero modes because the lattice breaks the translational invariance in this case, however the corresponding eigenvalues are small (see Fig. 4.9a, where the eigenvalues at A are shown as blue triangles). Mode 3 is the uniform breathing mode ($m = 0$) and is a low energy mode slightly above modes 1 and 2 in Fig. 4.9a. The next two modes directly above them correspond to elliptic distortions of the skyrmion shape ($m = \pm 2$) and the final two local modes are triangular distortions ($m = \pm 3$). Therefore, there are seven local states of the metastable well. The subsequent modes are part of the magnon continuum.

At the saddle points S_1 and S_2 of the collapse mechanisms (Fig. 4.10b and c), the breathing mode becomes unstable. At S_2, the translational modes are also unstable. In total, five local states exist at S_1, and three at S_2.

At the escape saddle point (Fig. 4.10d), the unstable skyrmion exhibits similar modes to that of its metastable counterpart, but they appear distorted by the presence of the edge. Six local modes are found. The unstable mode (mode 1) is a mode of translation towards the boundary, and is the mode that enables the escape of the skyrmion. The edge lifts the degeneracy of the translational modes: mode 2 is a mode of translation parallel to the boundary, and the associated eigenvalue is close to zero (see Fig. 4.9k, where the eigenvalues at S are shown as red circles) as this displacement costs little energy. Whether the boundary is flat or curved (Fig. 4.11a), the same number of internal modes exist at A and S, and they appear similar. There are, of course, fewer collective modes associated with the circular sample, since there are fewer magnetic sites and therefore fewer degrees of freedom in that case.

Next, Fig. 4.11b and c display examples of how the presence of a defect modifies some of the internal modes. In particular, the unstable breathing mode at the saddle point is significantly affected by the defect (Fig. 4.11c). The defect also lowers the total number of internal modes, as six internal modes survive at A, and four at S.

Lastly, in the system of two metastable coupled skyrmions, which we show in Fig. 4.11d, the internal modes are the same as that of a single skyrmion and both skyrmions are excited simultaneously, which results in a total of 16 internal modes. This is no longer the case at the transition state (Fig. 4.11e) where all the amplitude is localized to only one of the two skyrmions in each of the internal modes. Note that Fig. 4.11d and e show modes on the 50×50 lattice, whereas calculations were carried out on the 80×80 lattice to prevent the skyrmions from being constrainted by the edge, but the mode profiles looked similar.

4.4.2.3 Contribution of the Modes to the Prefactor

The internal modes that we have described above correspond, in fact, to the modes found below the magnon continuum in Fig. 4.9a–c, j–l. The separation between localized and collective modes is shown by a dashed line in Fig. 4.9, and also coincides with the transition between strong and low i-dependence of $\Omega_{0,i}$ (Fig. 4.9g–i, p–r). The contribution of the internal modes to the prefactor is given in Table 4.1 by $\Omega_{0,\text{int}}$, while the complete contribution of all the modes corresponds to $\Omega_{0,\text{tot}}$.

For the collapse mechanisms, the values differ by a factor of $\approx$40 between them, or 18 when a defect is present. In the case of the escape processes, the ratio is a factor of three (flat boundary), or four (curved boundary). In other words, the relative contribution of internal modes varies depending on the annihilation process, and is found to be higher for collapse processes compared to escape processes. It is reduced if a defect is present, and increases if the boundary is curved. Hence, the relative contribution of internal and collective modes is strongly linked to the nature of the annihilation and the geometry of the transition state. Most importantly, *internal modes play the most significant role in the thermally activated annihilation of a skyrmion*. Each of the collective spin-wave modes brings a weak contribution, but because there are many more collective modes than there are internal modes, their contribution to the attempt frequency should not be neglected.

4.4.2.4 Broken Symmetries

Finally, we note that the eigenmodes at the saddle point tend to display broken symmetries compared to the metastable states. In the single metastable skyrmion case (Fig. 4.10a), the internal modes in particular possess symmetries of types two-fold, four-fold, six-fold, and radial. At the saddle point of the collapse (Fig. 4.10b), the four-fold and six-fold symmetries are absent. As for the saddle configuration at the boundary (Figs. 4.10d and 4.11a), symmetries are broken by the edge. Lastly, in the case of the two skyrmions, the symmetry breaking at the SP in Fig. 4.11e is striking, since each internal mode involves only one skyrmion, in contrast to the metastable eigenmodes in Fig. 4.11d, which involve both skyrmions.

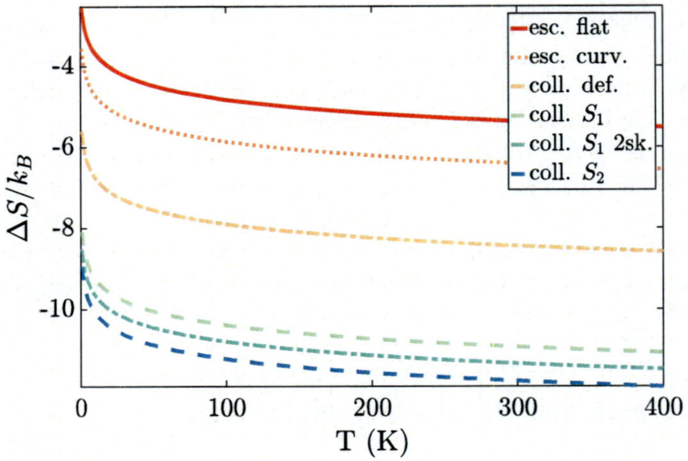

Fig. 4.12 Calculated change in configurational entropy induced when the system goes to the saddle point $\Delta S/k_B = (S_S - S_A)/k_B$ as defined in Eq. (3.38) over a broad range of temperatures for all mechanisms. The highest entropic barrier corresponds to the most negative ΔS, i.e. the collapse involving the saddle point S_2

4.4.3 Entropic Contribution to Skyrmion Stability

4.4.3.1 Variation of the Rate Between Different Paths to Annihilation

From the results shown in Table 4.1 and Fig. 4.7, a striking observation is that the most probable mechanisms do not necessarily correspond to the lowest energy barriers. For instance, for the small skyrmions considered here, escape mechanisms involve a higher barrier than collapse mechanisms, yet, at room temperature, escape is more probable. Therefore, *activation energies alone do not allow the lifetime of skyrmions to be properly estimated.* Characteristic times at the transition state were found to lie in the GHz-THz regime, which remains in the range of typically assumed values for estimates of the prefactor in magnetic spin systems [30–33]. However, a large difference is observed here due to the contribution of the ratio of curvatures, Ω_0, which significantly lowers the attempt frequency. As was explained in Sect. 3.2.2.2, Ω_0 characterizes the entropic contribution to the prefactor. By inverting Eq. (3.39), we can plot the change in configurational entropy, $\Delta S = S_S - S_A$, as a function of temperature. This is shown in Fig. 4.12 for all previously considered mechanisms. In all cases, we find $\Delta S < 0$: the configurational entropy of the metastable state is higher than that of the saddle point. This result implies that the number of micro-realisations of the metastable skyrmion state is higher than that of the transition state. A potential source of stability of individual skyrmions might therefore lie in lowered attempt frequencies due to entropic narrowing in the saddle point region, rather than in high internal energy barriers. The smaller reduction in entropy between A and S in the boundary annihilation may be explained by the fact that the transition state is a full skyrmion, which remains somewhat similar to the metastable skyrmion state. From Eq. (3.39), $\Omega_0(i)$ plotted in Fig. 4.9g–i, p–r in semilogarithmic scale behaves as $\Delta S(i)$. Hence, *the change in entropy is primarily associated with the internal modes of a skyrmion.*

The results in Table 4.1 and Fig. 4.7 show that at room-temperature, escape processes tend to be more probable than collapse, with the exception of the collapse on a defect. This occurs because, even though the energy barriers for collapse are lower, the attempt frequencies are also lower. While, at low temperature, the rate is dominated by the internal energy barrier, at higher temperature entropic effects contained in the attempt frequency are more important. If the attempt frequencies for all mechanisms were of a similar order of magnitude of a few gigahertz, collapse processes would remain the most probable mechanisms at all temperatures, and the average lifetime associated with collapse in the absence of defects would span between about 50 to 70 μs. In reality, it lies between 8 and 21 ms, which differs from the typically expected result by three orders of magnitude.

The escape through a curved boundary was found to be less likely than through a flat boundary, due to both the internal energy barrier and the entropic barrier being increased at the curved boundary. We also stress that the skyrmions we have considered so far only span over a few lattice sites and are stabilized at zero-field. For different stabilization processes involving an external field and lower perpendicular anisotropy, escape processes may be found more favourable also in terms of the internal energy barrier [10]. Interestingly, the collapse in the presence of another skyrmion exhibits the same internal energy barrier as in the case of a single skyrmion, but a larger stabilizing entropic contribution. The presence of a second skyrmion thus results in a lowered attempt frequency for collapse: $f_0^{\mathrm{col}} = 4.43$ MHz for the collapse via mechanism n°1 when a second skyrmion is present, as opposed to $f_0^{\mathrm{col}} = 6.70$ MHz for a single isolated skyrmion. This may be explained by the fact that the two skyrmion modes are coupled in the metastable state, and so 16 internal modes exist in total, which results in a larger configurational entropy in the metastable well, and consequently a lowered attempt frequency for collapse compared to the single skyrmion case. Finally, the presence of a non-magnetic defect significantly affects all terms of the transition rate. Firstly, as seen in Fig, 4.2c, it lowers the internal energy barrier by almost one order of magnitude, rendering it practically flat. It also decreases the entropic contribution to the prefactor by two orders of magnitude, that is, the activation entropoy is reduced (see also Fig. 4.12). Thirdly, the rate of growth of the unstable mode at the saddle point is decreased by one order of magnitude. This effect could, in theory, be stabilizing, but we find it is negligible against the significantly lower value of the total activation energy. It can also be noted that the internal energy barrier for the nucleation is lowered as well. This is consistent with experimental observations that skyrmions tend to nucleate and annihilate near defects [9].

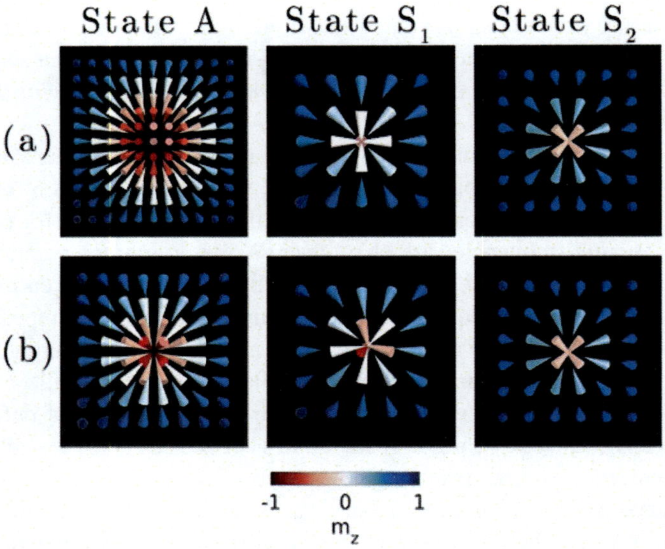

Fig. 4.13 Spin maps (zoomed) of the metastable state A and saddle points S_1 and S_2 of the skyrmion collapse for **a** $b_z = 0$, and **b** $b_z = 0.05$. S_1 corresponds to the skyrmion core centred on a lattice site, while S_2 corresponds to the core located at an interstitial point

4.4.3.2 Variation of the Collapse Rate as a Function of Applied Field

We now consider the effect of an external magnetic field applied along the $+z$ direction, so that it points opposite to the core of the skyrmion and therefore has a destabilizing effect. We focus on the collapse processes of a single skyrmion. For reduced field values $b_z \in [0, 0.05]$, we relax the corresponding collapse mechanisms n°1 and 2. With our present set of parameters, this corresponds to a field variation from 0 to about 730 mT. The relaxed metastable skyrmion core may either be centred on a site (State A in Fig. 4.13a), or at an interstitial position (State A in Fig. 4.13b). This is determined by its equilibrium size and its commensurability with the underlying lattice. If, for a given field value, we attempt to relax the skyrmion at a different position on the lattice—i.e., centred on a site if it initially relaxed at an interstitial point, and vice versa—it will relax back to its initial position. Therefore, there seems to exist only one type of stable skyrmion state A per field value, although if we apply periodic boundary conditions, the skyrmion may of course be indistinguishably located at either of the N possible sites. If the metastable skyrmion is pinned at an interstitial point, the realisation of mechanism n°1 requires the core to firstly shift onto a lattice site. For all field values, S_2 is found to be of order three. With this in mind, the rate associated with mechanism n°2 will contain some larger degree of error (see Table A.2), as we are not able to relax the CI-GNEB force below a certain tolerance. That tolerance varies depending on the applied field strength ($9 \times 10^{-7} < F_{\text{tol}} < 6 \times 10^{-5}$).

As we increase the field, the stable skyrmion size decreases—in agreement with theoretical and experimental studies [34–36]—and so does the internal energy barrier for both collapse mechanisms, which we show in Fig. 4.14a. The energy barriers associated with S_1 and S_2 are almost identical and decrease smoothly with the applied field. Note that the barrier at S_2 is slightly higher than at S_1.

In Fig. 4.14b, we plot the change in entropy, $\Delta S_{1,2} = S_{S_{1,2}} - S_A$, respectively associated to S_1 and S_2, as defined in Eq. (3.38) for a first-order saddle point such as S_1, and more generally in Eq. (3.40) for a kth-order saddle point. The latter is relevant for S_2, for which $k = 3$. $\Delta S_{1,2}$ is found to increase (become less negative) as the field increases. Since $\Delta S_{1,2} < 0$ for all field values, the entropic contribution remains stabilizing, as it lowers the attempt frequency, but this stabilizing contribution decreases at higher field values. We also find that $\Delta S_2 < \Delta S_1 < 0$, so $S_{S_2} < S_{S_1}$: statistically, S_1 is more likely than S_2 to be visited by the system under thermal fluctuations.

In Fig. 4.14c, we show the value of the dynamical contribution to the prefactor at saddle points S_1 and S_2, namely $\lambda_{+,1}$ and $\lambda_{+,2}$, as a function of the applied field. We find that they decrease weakly when the field increases. $\lambda_{+,2}$ is slightly larger than $\lambda_{+,1}$ and decreases more than $\lambda_{+,1}$. For both of them, we consistently obtain a value in the range of a few terahertz, with little variation as a function of the applied field, i.e., $0.8 < \lambda_{+,1} < 1.2$ THz and $1.15 < \lambda_{+,2} < 2.12$ THz. Following the discussion in Sect. 3.3.2, since S_2 is a third-order saddle point, $\lambda_{+,2}$ is chosen as the rate of growth of the fastest growing instability [37], which corresponds to the unstable breathing mode.

If we assume that collapse mechanisms n° 1 and 2 are independent, the total rate of collapse of the skyrmion can be expressed as

$$k_{\text{tot}}^{\text{col}}(T) = k_1(T) + k_2(T) = f_{01}e^{-\beta \Delta E_1} + f_{02}e^{-\beta \Delta E_2}, \tag{4.2}$$

where f_{01} and f_{02} are the attempt frequencies respectively associated with mechanisms n°1 and 2. We should note that since the difference in internal energy between S_1 and S_2 is quite small (around 0.08 J_{ex} at zero field), the mechansisms are probably not completely decoupled. In Fig. 4.14d, we plot the total collapse rate calculated with Eq. (4.2), as well as the individual rates k_1 and k_2. For each field value, we set the value of T such that $\beta \Delta E_1 = 10$. We find that k_1 is much larger than k_2, so that $k_1 + k_2 \approx k_1$, despite the fact that the internal energy barriers are almost identical. In the higher field region ($0.04 \leq b_z \leq 0.05$), the configuration at S_1 loses cubic symmetry (i.e., invariance under $\pi/2$-rotations of the system about the z-axis) as seen in Fig. 4.13b, and possesses four equivalent realisations. They can be visualized by picturing rigid $\pi/2$—rotations of the system. However, the non-symmetric state S_2 corresponds to a small deviation δ from the symmetric state. If the amplitude of the thermal fluctuations exceeds δ, then the four states cannot be resolved at finite temperature and, on average, should correspond to a single state. For this reason, we do not include a factor of four in the rate. This choice shall be validated in Chap. 5.

Lastly, since the energy barriers involved in both mechanisms are very similar (Fig. 4.14a), we can recover an approximate total attempt frequency for the collapse as

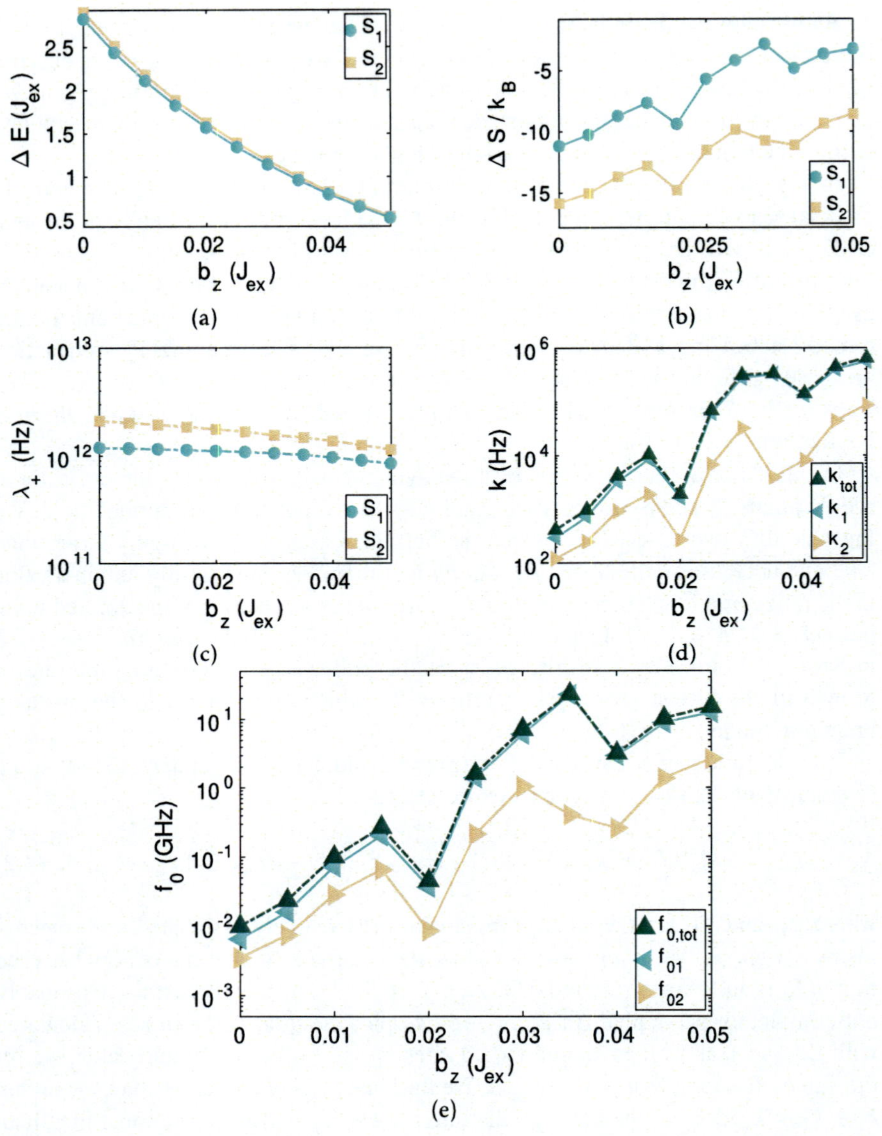

Fig. 4.14 As a function of the reduced destabilizing field b_z and for collapse mechanisms n°1 and 2, we show the following quantities. **a** Internal energy barriers. They both decrease smoothly with increasing field. We find $\Delta E_1 \lessapprox \Delta E_2$. **b** Change in configurational entropy when the system goes from states A to $S_{1,2}$ at $T = 300$ K. We find $\Delta S_2 < \Delta S_1 < 0$. Since the change is negative in both cases, the entropy decreases at the SP, which is a stabilizing effect. The non-monotonic variations were attributed to lattice effects. **c** Dynamical contribution to the prefactor. For both mechanisms, it decreases only weakly with increasing field, with $\lambda_{+,1} \lessapprox \lambda_{+,2} \approx$ THz. **d** Total rate of collapse at a temperature such that $\beta \Delta E_1 = 10$ calculated from Eq. (4.2), where $k_2(T) \ll k_1(T)$ despite the internal energy barriers being almost identical. **e** Attempt frequency for the collapse. Since $\Delta S_2 < \Delta S_1 < 0$, we have $f_{0,2} \ll f_{0,1}$, which in turns explain why $k_2(T) \ll k_1(T)$

$f_{0,\text{tot}}^{\text{col}} \approx f_{01} + f_{02}$, where the energy barrier is approximated as $\Delta E_1 \approx \Delta E_2 \approx \Delta \bar{E}$, with

$$\Delta \bar{E} = \frac{\Delta E_1 + \Delta E_2}{2}. \tag{4.3}$$

The attempt frequencies f_{01}, f_{02}, and $f_{0,\text{tot}}^{\text{col}}$ are plotted in Fig. 4.14e as a function of the reduced field. $f_{0,\text{tot}}^{\text{col}}$ spans over three orders of magnitude, that is, 10.2 MHz $\leq f_{0,\text{tot}}^{\text{col}} \leq$ 23.24 GHz. In particular, 6.7 MHz $\leq f_{01} \leq$ 22.8 GHz, and 3.5 MHz $\leq f_{02} \leq$ 2.68 GHz. This shows a strong dependence of the attempt frequency on the applied magnetic field. The dynamical contribution is bound in the THz region and does not explain why the attempt frequency extends from the MHz to the GHz regime. This effect was also been discussed in Refs. [7, 12] and once again finds its origin in the strong entropic contribution to the prefactor, as shown in Fig. 4.14b. Additionally, we typically obtain $f_{0,2} \ll f_{0,1}$, which in turns explains why the total rate $k_{\text{tot}}^{\text{col}}$ is dominated by k_1 even though $\Delta E_1 \approx \Delta E_2$. Once again, since $\lambda_{+,1} < \lambda_{+,2}$, this is a direct consequence of the different entropic contributions, more specifically $\Delta S_2 < \Delta S_1 < 0$. This example hints that considering only first-order saddle points when applying some form of reaction rate theory is probably justified. We note that, since the skyrmions are not decoupled from the lattice, we observe strong lattice effects that result in non-monotonic variations of the entropic contribution (Fig. 4.14b), which are in turn found in the variations of the attempt frequency (Fig. 4.14e) and the total collapse rate (Fig. 4.14d).

4.5 Discussion

In this chapter, we applied Langer's theory to compute annihilation rates of metastable chiral magnetic skyrmions with respect to collapse and escape processes. Two separate mechanisms were identified for the collapse of a single skyrmion. They involve different saddle points, and so each of them is associated with its own rate. We identified the thermally significant modes as the skyrmion's internal modes, while the other modes pertain to collective fluctuations that can be interpreted as spin-wave excitations, which contribute weakly. By changing the underlying symmetry of the DMI, we were able to check that the present results hold for not only Néel skyrmions, but also Bloch skyrmions and antiskyrmions. We found that collapse on a non-magnetic defect yields the shortest mean waiting times between similar events, with $\tau \approx 0.1\,\mu$s at room temperature. In the absence of defects, collapse mechanisms were found more probable than escape at low temperature, but this trend reverses at room temperature, at which escape processes are more probable. The largest waiting time at 300 K was found to be 21 ms, and is associated with collapse mechanism n°2. The presence of a second skyrmion was shown to decrease the probability of collapse. Finally, escape through a flat boundary was found more probable than through a curved boundary.

We used a simple Heisenberg model limited to first-neighbour exchange interactions and no dipole-dipole couplings, but which should nevertheless capture the essential physics behind skyrmion annihilations in ultrathin films. Similar stability studies where the exchange was parametrized by DFT calculations and went beyond the first nearest-neighbour approximation reached very similar conclusions concerning the importance of the prefactor [7, 38]. In systems were dipole-dipole interactions were found to play an important role in the skyrmions' stability, it was also demonstrated that an effective anisotropy is enough to reproduce similar energy barriers [39]. While this is valid in two-dimensional systems, it may however not be the case in three dimensions. Moreover, the effect of dipolar interactions on the attempt frequency has not been studied, since dipolar couplings yield non-sparse matrices which are difficult to diagonalize in large systems.

The important result of this chapter is the strong entropic contribution to skyrmion stability. In the configurations studied here, the configurational entropy decreases between the metastable state and the saddle point, which results in lowered attempt frequencies. This therefore constitutes a stabilizing effect, which we have referred to as entropic narrowing in the saddle region. The strong entropic contribution is primarily associated with the skyrmion's internal modes, and is more pronounced for collapse mechanisms. The aforementioned effect has two main consequences.

i. When considering different types of processes, the rates of annihilation, $k(T)$, are not necessarily ordered like the internal energy barriers. A mechanism with a higher energy barrier might turn out to be more probable because it possesses a low attempt frequency. One example is collapse and escape mechanisms at zero-field. In the present case, collapse processes have a lower energy barrier than escape, but also a much lower attempt frequency. As a consequence, escape processes are more probable than collapse at room temperature. Another example is the case of collapse mechanisms n° 1 and 2. They have almost the same internal energy barriers, but the attempt frequency associated with mechanism n°2 is much lower, so the total rate is dominated by mechanism n°1.
ii. The attempt frequency, and therefore the rate, displays a strong dependence on external parameters, such as external magnetic field.

For these reasons, internal energy barriers are not enough to correctly predict the lifetime of skyrmions, and it is essential to also evaluate a rate prefactor.

In principle, the total rate of annihilation of an isolated skyrmion in a finite-size system is given by the sum of the rates due to collapse and escape through a boundary, $k^{\text{tot}} = k^{\text{col}} + k^{\text{esc}}$. However, the escape rates we compute here through Langer's theory seem to be instantaneous rates. Since they do not depend on the size of the system (see Table A.1), they do not take into account the time it takes for the skyrmion to diffuse to the boundary. In Ref. [7], the skyrmions are decoupled from the lattice in the metastable state, which results in the emergence of translational Goldstone modes, and a system-size dependence of the rate. This is not satisfactory either, since the breaking of the translational invariance would also result in the loss of system-size dependence. The escape rate we compute therefore corresponds to the probability that a skyrmion which is found close to the boundary, and which cannot diffuse back inside the sample, will escape. The Brownian motion performed by an isolated skyrmion under thermal fluctuations was shown to be diffusive, i.e., with a linear time-dependence of the mean square displacement [14],

$$\langle \Delta \mathbf{R}^2 \rangle = \langle \big(\mathbf{R}(t) - \mathbf{R}(t=0)\big)^2 \rangle \sim \mathcal{D}_{\text{sk}} t, \tag{4.4}$$

where $\mathbf{R}(t)$ is the position of the skyrmion at time t, and $\mathcal{D}_{\text{sk}}$ is the diffusion coefficient of the skyrmion as a particle. The diffusion time Δt_{diff} to reach the boundary at a distance R_b from the initial skyrmion position should in principle behave as $\Delta t_{\text{diff}} \sim R_b^2/\mathcal{D}_{\text{sk}}$, so the full probability that the skyrmion may escape through the boundary should also depend on T, R_b^2, etc., without the need for translational invariance.

Lastly, the results in this chapter were obtained by applying Langer's theory to the problem of skyrmion lifetime. As we have seen in Chap. 3, this theory is based on many assumptions concerning the damping regime, the height of the barrier, and a state of quasi-thermal equilibrium of the system during the transition. Moreover, whenever we are faced with several mechanisms for a single process, as is the case of the skyrmion collapse, we can only assume the mechanisms are independent in order for the rates to add up (e.g., Eq. (4.2)), which may not hold. We are also faced with the existence of higher-order saddle points, and the question of whether they should contribute to the total rate. An alternative approach is therefore required in order to validate the present results. This will be explored in the next chapter.

References

1. Desplat L, Suess D, Kim J-V, Stamps RL (2018) Thermal stability of metastable magnetic skyrmions: entropic narrowing and significance of internal eigenmodes. Phys Rev B 98:134407
2. Desplat L, Vogler C, Kim J-V, Stamps RL, Suess D (2020) Path sampling for lifetimes of metastable magnetic skyrmions and direct comparison with Kramers' method. Phys Rev B 101(6):060403(R)
3. Ritzmann U, Von Malottki S, Kim J-V, Heinze S, Sinova J, Dupé B (2018) Trochoidal motion and pair generation in skyrmion and antiskyrmion dynamics under spin-orbit torques. Nat Electron 1(8):451
4. Thiaville A, Rohart S, Jué É, Cros V, Fert A (2012) Dynamics of Dzyaloshinskii domain walls in ultrathin magnetic films. Europhys Lett 100(5):57002
5. Heo C, Kiselev NS, Nandy AK, Blügel S, Rasing T (2016) Switching of chiral magnetic skyrmions by picosecond magnetic field pulses via transient topological states. Sci Rep 6:27146
6. Bessarab PF, Uzdin VM, Jonsson H (2015) Method for finding mechanism and activation energy of magnetic transitions, applied to skyrmions and antivortex annihilation. Comput Phys Commun 196:335–347
7. Bessarab PF, Müller GP, Lobanov IS, Rybakov FN, Kiselev NS, Jónsson H, Uzdin VM, Blügel S, Bergqvist L, Delin A (2018) Lifetime of racetrack skyrmions. Sci Rep 8(1):1–10
8. Romming N, Hanneken C, Menzel M, Bickel JE, Wolter B, von Bergmann K, Kubetzka A, Wiesendanger R (2013) Writing and deleting single magnetic skyrmions. Science 341(6146):636–639
9. Hanneken C, Kubetzka A, Von Bergmann K, Wiesendanger R (2016) Pinning and movement of individual nanoscale magnetic skyrmions via defects. New J Phys 18(5):055009

10. Uzdin VM, Potkina MN, Lobanov IS, Bessarab PF, Jónsson H (2017) The effect of confinement and defects on the thermal stability of skyrmions. Phys B: Condens Matter 549:6–9
11. Meyer S, Perini M, von Malottki S, Kubetzka A, Wiesendanger R, von Bergmann K, Heinze S (2019) Isolated zero field sub-10 nm skyrmions in ultrathin Co films. Nat Commun 10:3823
12. Wild J, Meier TN, Pöllath S, Kronseder M, Bauer A, Chacon A, Halder M, Schowalter M, Rosenauer A, Zweck J et al (2017) Entropy-limited topological protection of skyrmions. Sci Adv 3(9):e1701704
13. Müller GP, Bessarab PF, Vlasov SM, Lux F, Kiselev NS, Blügel S, Uzdin VM, Jónsson H (2018) Duplication, collapse, and escape of magnetic skyrmions revealed using a systematic saddle point search method. Phys Rev Lett 121(19):197202
14. Schütte C, Iwasaki J, Rosch A, Nagaosa N (2014) Inertia, diffusion, and dynamics of a driven skyrmion. Phys Rev B 90(17):174434
15. Lin S-Z, Hayami S (2016) Ginzburg-Landau theory for skyrmions in inversion-symmetric magnets with competing interactions. Phys Rev B 93(6):064430
16. Rohart S, Thiaville A (2013) Skyrmion confinement in ultrathin film nanostructures in the presence of Dzyaloshinskii-Moriya interaction. Phys Rev B 88(18):184422
17. Sampaio J, Cros V, Rohart S, Thiaville A, Fert A (2013) Nucleation, stability and current-induced motion of isolated magnetic skyrmions in nanostructures. Nat Nanotechnol 8(11):839
18. Fiedler G, Fidler J, Lee J, Schrefl T, Stamps RL, Braun H, Suess D (2012) Direct calculation of the attempt frequency of magnetic structures using the finite element method. J Appl Phys 111(9):093917
19. Makhfudz I, Krüger B, Tchernyshyov O (2012) Inertia and chiral edge modes of a skyrmion magnetic bubble. Phys Rev Lett 109(21):217201
20. Lin S-Z, Batista CD, Saxena A (2014) Internal modes of a skyrmion in the ferromagnetic state of chiral magnets. Phys Rev B 89(2):024415
21. Schütte C, Garst M (2014) Magnon-skyrmion scattering in chiral magnets. Phys Rev B 90(9):094423
22. Iwasaki J, Beekman AJ, Nagaosa N (2014) Theory of magnon-skyrmion scattering in chiral magnets. Phys Rev B 89(6):064412
23. Buijnsters F, Fasolino A, Katsnelson M (2014) Zero modes in magnetic systems: general theory and an efficient computational scheme. Phys Rev B 89(17):174433
24. Schroeter S, Garst M (2015) Scattering of high-energy magnons off a magnetic skyrmion. Low Temp Phys 41(10):817–825
25. Zhang VL, Hou CG, Di K, Lim HS, Ng SC, Pollard SD, Yang H, Kuok MH (2017) Eigenmodes of Néel skyrmions in ultrathin magnetic films. AIP Adv 7(5):055212
26. Mruczkiewicz M, Krawczyk M, Guslienko KY (2017) Spin excitation spectrum in a magnetic nanodot with continuous transitions between the vortex, Bloch-type skyrmion, and Néel-type skyrmion states. Phys Rev B 95(9):094414
27. Guslienko KY, Gareeva ZV (2017) Gyrotropic skyrmion modes in ultrathin magnetic circular dots. IEEE Magn Lett 8:1–5
28. Garst M, Waizner J, Grundler D (2017) Collective spin excitations of helices and magnetic skyrmions: review and perspectives of magnonics in non-centrosymmetric magnets. J Phys D: Appl Phys 50(29):293002
29. Kravchuk VP, Sheka DD, Rößler UK, van den Brink J, Gaididei Y (2018) Spin eigenmodes of magnetic skyrmions and the problem of the effective skyrmion mass. Phys Rev B 97(6):064403
30. Weller D, Moser A (1999) Thermal effect limits in ultrahigh-density magnetic recording. IEEE Trans Magn 35(6):4423–4439
31. Chen E, Apalkov D, Diao Z, Driskill-Smith A, Druist D, Lottis D, Nikitin V, Tang X, Watts S, Wang S et al (2010) Advances and future prospects of spin-transfer torque random access memory. IEEE Trans Magn 46(6):1873–1878
32. Lederman M, Schultz S, Ozaki M (1994) Measurement of the dynamics of the magnetization reversal in individual single-domain ferromagnetic particles. Phys Rev Lett 73(14):1986
33. Cortés-Ortuño D, Wang W, Beg M, Pepper RA, Bisotti M-A, Carey R, Vousden M, Kluyver T, Hovorka O, Fangohr H (2017) Thermal stability and topological protection of skyrmions in nanotracks. Sci Rep 7(1):4060

34. Bogdanov A, Hubert A (1994) The properties of isolated magnetic vortices. Physica Status Solidi B 186(2):527–543
35. Wilson MN, Butenko AB, Bogdanov AN, Monchesky TL (2014) Chiral skyrmions in cubic helimagnet films: the role of uniaxial anisotropy. Phys Rev B 89:094411 Mar
36. Romming N, Kubetzka A, Hanneken C, von Bergmann K, Wiesendanger R (2015) Field-dependent size and shape of single magnetic skyrmions. Phys Rev Lett 114(17):177203
37. Duff G (2008) Langer's method for the calculation of escape rates and its application to systems of ferromagnets. PhD thesis, Dublin Institute of Technology
38. von Malottki S, Bessarab PF, Haldar S, Delin A, Heinze S (2019) Skyrmion lifetime in ultrathin films. Phys Rev B 99(6):060409
39. Lobanov IS, Jónsson H, Uzdin VM (2016) Mechanism and activation energy of magnetic skyrmion annihilation obtained from minimum energy path calculations. Phys Rev B 94(17):174418

Chapter 5
Skyrmion Collapse Rate Computation via Forward Flux Sampling and Comparison with Langer's Theory

5.1 Introduction

Generally speaking, the problem of annihilation rates of skyrmions pertains to the realm of rare events. That is, in a direct Langevin dynamics simulation, few—if any—annihilation events are observed. That is because the small timesteps [1] required to properly resolve the precessional dynamics of magnetic spin systems entails that direct simulations are, in practice, limited to a few hundred nanoseconds. Since skyrmions are required to be stable on the scale of ten years at room temperature for data storage applications, a better method than brute force simulations is required. In this chapter, we demonstrate the use of a path sampling method for the simulation of rare events, namely the forward flux sampling method, to compute collapse rates of magnetic skyrmions. We compare the obtained results with the ones calculated in Chap. 4 from the application of Langer's theory. The results presented in this chapter are based on an original publication [2].

The chapter is organized as follows. In Sect. 5.2, we begin by exposing the principles of Langevin spin dynamics and its implementation. In Sect. 5.3, we present the forward flux sampling method, which we then apply in Sect. 5.4 to compute collapse rates of isolated skyrmions. The results are compared with the ones previously obtained in Chap. 4 from the application of Langer's theory and are, lastly, discussed in Sect. 5.5.

5.2 Langevin Dynamics

5.2.1 Principle

A prerequisite for the implementation of the forward flux sampling method is the ability to properly solve the Brownian dynamics of the magnetic spin system at finite

L. Desplat, *Thermal Stability of Metastable Magnetic Skyrmions*, Springer Theses,
https://doi.org/10.1007/978-3-030-66026-0_5

temperature T. At zero temperature, the dynamics of each spin $\mathbf{m}_i$ is governed by the LLG equation (Eq. (3.46) and Fig. 5.1a), which, at finite temperature, may take the form of a Langevin equation (Eq. (3.4)) through the introduction of a stochastic magnetic field, as given in Ref. [3]

$$\frac{\mathrm{d}\mathbf{m}_i}{\mathrm{d}\bar{t}} = \frac{1}{\alpha}\mathbf{m}_i \times \left[\mathbf{b}^i_{\mathrm{eff}} + \mathbf{b}^i_{\mathrm{fl}}(\bar{t})\right] - \mathbf{m}_i \times \left(\mathbf{m}_i \times \left[\mathbf{b}^i_{\mathrm{eff}} + \mathbf{b}^i_{\mathrm{fl}}(\bar{t})\right]\right). \quad (5.1)$$

Equation (5.1) is the dimensionless stochastic LLG equation. As in previous chapters, the quantities are in units of the isotropic exchange coupling constant between first nearest-neighbours, $J_1 = J_{\mathrm{ex}}$, such that in the above expression, $\mathbf{b}^i_{\mathrm{eff}}$ is the reduced local effective field, as defined in Eq. (3.80), and $\mathbf{b}^i_{\mathrm{fl}}$ is a stochastic fluctuating field in the form of white noise, which accounts for fluctuations of the orientation of $\mathbf{m}_i$ caused by interactions with microscopic degrees of freedom of the environment such as phonons, conducting electrons, nuclear spins, etc. It is assumed to be Gaussian distributed, with the following statistical properties, in agreement with the fluctuation-dissipation theorem,

$$\langle b^i_l(\bar{t})\rangle = 0, \qquad \langle b^i_l(\bar{t}) b^j_m(\bar{t}')\rangle = 2\mathcal{D}\delta_{lm}\delta_{ij}\delta(\bar{t} - \bar{t}'), \quad (5.2)$$

where l, m are Cartesian indices, i, j are site indices, and

$$\mathcal{D} = \frac{\alpha^2}{1+\alpha^2}\frac{k_B T}{J_1} \quad (5.3)$$

is the reduced diffusion coefficient. $\langle \ldots \rangle$ denotes an average over different realizations of the fluctuating field (see Sect. 3.1.1). The Dirac distribution, $\delta(\bar{t} - \bar{t}')$, translates the fact that, above a certain temperature, the autocorrelation times of $\mathbf{b}^i_{\mathrm{fl}}$ is much shorter than the rotational response of the system. The Kronecker symbol δ_{lm} comes from the assumption that the different components of $\mathbf{b}^i_{\mathrm{fl}}$ are uncorrelated, while δ_{ij} means that the fluctuating fields at different sites are uncorrelated. The reduced time $\bar{t}$ is linked to physical time t via

$$t = \frac{M_s}{\alpha\gamma_e J_1}\bar{t}. \quad (5.4)$$

In Fig. 5.1b, we show an example of the precession of a single spin about the deterministic effective field obtained by solving Eq. (5.1) at finite temperature. The inclusion of the stochastic fluctuating field results in a stochastic trajectory. As the stochastic field enters in a multiplicative way, an interpretation rule is needed in order to properly define it. This is because the noise has zero correlation time (Eq. (5.2)), so $\mathbf{m}_i$ will jump at every time the noise is realized, and so it is not clear which value of $\mathbf{m}_i$ should enter on the RHS of Eq. (5.1). Since the white noise is an idealisation of physical noise with short autocorrelation time, the Stratonovich interpretation is the better choice in this case [4]. The Dirac $\delta(t)$ may be approximated by a function

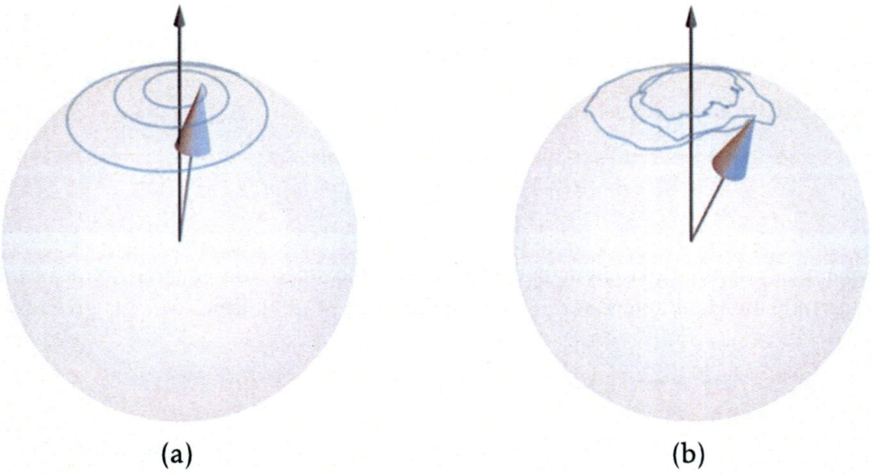

Fig. 5.1 Damped precession of a single magnetic spin (coloured arrow) about the effective magnetic field (dark arrow) **a** at zero temperature (deterministic LLG), **b** at finite temperature (stochastic LLG). In stochastic LLG, the coupling to the environment is modelled by a stochastic fluctuating magnetic field in the form of white noise

$\delta_\epsilon(t)$ of very small width ϵ, as [5]

$$\delta_\epsilon(t) = \begin{cases} \frac{1}{\epsilon} & \text{if } -\frac{\epsilon}{2} < t < \frac{\epsilon}{2}, \\ 0 & \text{otherwise.} \end{cases} \tag{5.5}$$

One then takes the limit $\epsilon \to 0$ in the final result.

5.2.2 *Numerical Implementation*

5.2.2.1 The Stochastic Heun Scheme

For convenience, we drop the i-index associated with magnetic sites, as well as the overbar on the reduced time in this section. The system of stochastic dimensionless LLG equations (Eq. (5.1)) at each spin is cast into a system of Langevin equations (Eq. (3.4)) along each Cartesian axis $l, m, n = \{1, 2, 3\}$ as [1]

$$\frac{\mathrm{d}m_l}{\mathrm{d}t} = A_l(\boldsymbol{M}, t) + \sum_n B_{ln}(\boldsymbol{M}, t) b_{\mathrm{fl},n}(t), \tag{5.6}$$

in which $\boldsymbol{M} = [\mathbf{m}_1, \ldots \mathbf{m}_N]$. In the previous expression,

$$A_l = \sum_n B_{ln} b_{\text{eff},n} \tag{5.7}$$

is the deterministic right-hand side of Eq. (5.1), and

$$B_{ln} = \frac{1}{\alpha} \sum_m \epsilon_{lmn} m_m + (\delta_{ln} - m_l m_n), \tag{5.8}$$

where ϵ_{lmn} is the Levi-Civita symbol. We use the stochastic generalization of the deterministic Heun integrator [1], where the value of m_l at time $t + dt$ is given by

$$\begin{aligned} m_l(t+dt) = m_l(t) &+ \frac{1}{2}\Big[A_l(\tilde{\boldsymbol{M}}, t+dt) + A_l(\boldsymbol{M}, t)\Big]dt \\ &+ \frac{1}{2}\sum_n \Big[B_{ln}(\tilde{\boldsymbol{M}}, t+dt) + B_{ln}(\boldsymbol{M}, t)\Big]\Delta W_n(t). \end{aligned} \tag{5.9}$$

In the above expression, dt is the integration time step, and $\tilde{\boldsymbol{M}} = [\tilde{\mathbf{m}}_1, \ldots, \tilde{\mathbf{m}}_N]$ are the Euler-type supporting values, with each Cartesian component $\tilde{m}_{l,i} = \tilde{m}_l$ of $\tilde{\mathbf{m}}_i$ expressed as

$$\tilde{m}_l = m_l(t) + A_l(\boldsymbol{M}, t)dt + \sum_n B_{ln}(\boldsymbol{M}, t)\Delta W_n. \tag{5.10}$$

Lastly,

$$\Delta W_n = \int_t^{t+dt} \mathrm{d}t' b_{\mathrm{fl},n}(t') \tag{5.11}$$

are Gaussian distributed random numbers of first and second moments

$$\langle \Delta W_n \rangle = 0, \qquad \langle \Delta W_l \Delta W_m \rangle = 2\mathcal{D} dt \delta_{lm}, \tag{5.12}$$

where $\mathcal{D}$ is defined by Eq. (5.3). The scheme converges to the solution of (5.6) when interpreted in the sense of Stratonovich.

5.2.2.2 Implementation

The stochastic Heun integrator for Eq. (5.1) was implemented by the author in C++ following Ref. [1]. At each time step, the Cartesian components of the stochastic field are drawn from a normal distribution defined by Eq. (5.12) using the Mersenne Twister 19937 pseudo-random number generator (MT19937) [6].

Table 5.1 Theoretical and numerical time average of the component of the magnetization parallel to the anisotropy axis for different temperatures at thermal equilibrium

Temperature (K)	$\langle m_z \rangle_{t,\mathrm{th}}$	$\langle \lvert m_z \rvert \rangle_{t,\mathrm{num}}$
10	0.98979	0.98978
50	0.94268	0.94282
200	0.71976	0.71686

5.2.3 Numerical Tests

5.2.3.1 Single Macrospin

Based on the tests carried out in Ref. [7], we simulate a single rigid magnetic moment in the presence of an easy-axis anisotropy along z, and the following system parameters taken from Ref. [7]: $\mathscr{M}_s = 1.3$ MA/m; $\mathscr{K} = 6.9$ MJ/m^3; $\alpha = 0.1$; $a = 1$ nm. The reduced time step was set to $\mathrm{d}\bar{t} = 0.001$, which corresponds to a physical time step of $\mathrm{d}t = 4.55 \times 10^{-15}$ s. The first test concerns the correct resolution of the deterministic precessional dynamics of the single-spin system. We initialize the simulation with a tilt angle $\theta = \pi/4$ with respect to the z-axis, and we let the magnetization relax along the anisotropy axis at $T = 0$ K. The resulting dynamics is shown in Fig. 5.1a. In Fig. 5.2a we plot the time evolution of the Cartesian components of the single spin. The corresponding theoretical precession period is [7]

$$\mathcal{T}_{\mathrm{th}} = \frac{2\pi}{\gamma_e B_{\mathrm{eff}}} = 3.32 \text{ ps}. \tag{5.13}$$

We measure the time interval between five maxima of m_y in Fig. 5.2a and obtain $\mathcal{T}_{\mathrm{num}} = 3.39$ ps, which shows good agreement with the theoretical value.

The second test aims to validate the implementation of the stochastic fluctuating field. The theoretical time average for the z-component of the magnetization at thermal equilibrium, $\langle m_z \rangle_{t,\mathrm{th}}$, is given in Table 5.1 [7]. We compute $\langle m_z \rangle_{t,\mathrm{num}}$ over a time period $\Delta t = 10^8\, dt$, starting from $t_0 = 10^6\, dt$ in order to allow the system to thermalize. Magnetization reversal events may occur, which are not taken into account in the theoretical average. We thus compute $\langle \lvert m_z \rvert \rangle_{t,\mathrm{num}}$ to reduce the error due to reversals. However, at $T = 200$ K, reversals are rather frequent, and the numerical average is thus slightly under-evaluated. The results are gathered in the last column of Table 5.1 and plotted in Fig. 5.2b, where the theoretical value is shown as dotted lines. We obtain a good agreement with the theory.

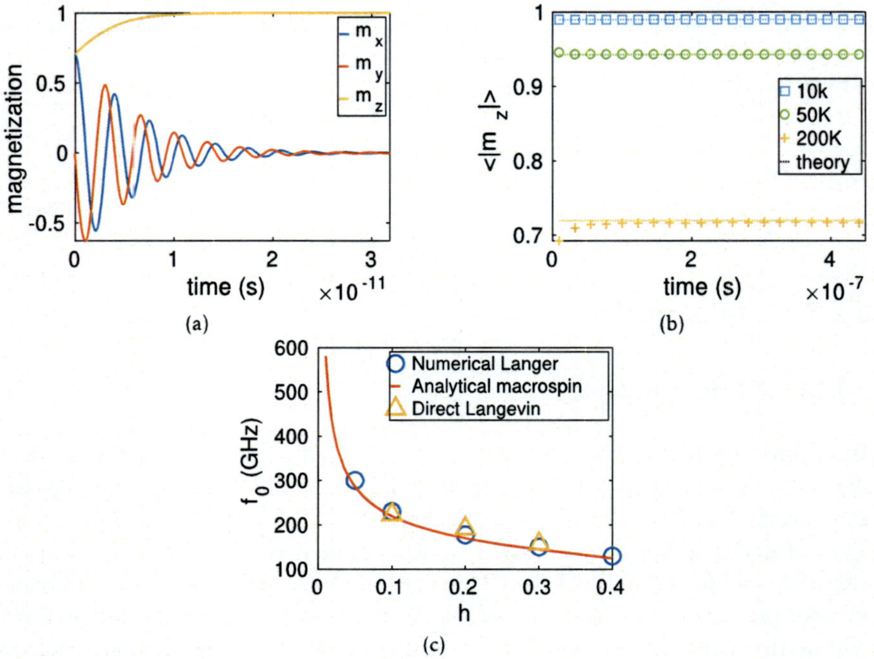

Fig. 5.2 **a** Precessional relaxation of the magnetization for the single rigid magnetic moment at $T = 0$ K, **b** numerical and theoretical time average value of $|m_z|$ calculated after thermalization for different temperatures, **c** attempt frequency of coherent magnetization reversal in a perpendicular applied field, comparison between the analytical result from Langer's theory for a single macrospin (Eq. (3.94)), and, for $N = 7 \times 7$ spins, Langer's theory implementation and direct Langevin simulations at a temperature which verifies $\beta \Delta E = 5$

5.2.3.2 Magnetization Reversal in a Perpendicular Applied Field

Finally, we go back to the case of coherent magnetization reversal under a perpendicular applied field treated in Sect. 3.5.4. We now aim to compare the attempt frequency obtained from direct Langevin simulations with previous results from Langer's theory. For three arbitrary values of the reduced field h, as defined in Eq. (3.93), we set the temperature T such that $\beta \Delta E = 5$, where ΔE is the activation barrier for the reversal. The system parameters are the same as in Sect. 3.5.4, but we now simulate a square system of $N = 7 \times 7$ spins. In that case, the numerical result of Langer's theory begins to deviate slightly from the theoretical macrospin case, due to the contribution of collective modes. This can be seen in Fig. 5.2c, where the values of f_0 computed with the numerical implementation of Langer's theory (shown as blue circles) do not match exactly with the theoretical formula (shown in red). The system is initialized in state A with $\phi_{A,i} = \arcsin(h)\ \forall i$. We launch 400 Langevin simulations and compute an average waiting time for the reversal—i.e., the average time it takes for the system to reach state B, which corresponds to $\phi_{B,i} = \pi - \arcsin(h)$.

We consider a reversal has happened if [8]

$$|\mathbf{M}(t) - \mathbf{M}_B| < \lambda_B, \tag{5.14}$$

where $\mathbf{M}_B = [\mathbf{m}_{B,1}, \ldots \mathbf{m}_{B,N}]$, the vertical bars denote the (N-dimensional) Euclidian norm, and λ_B defines the boundary of region B. λ_B must be chosen such that it does not impact the value of the reversal rate k_{AB}. According to [8], $0 < \lambda_B \leq 0.5$ is within acceptable values, so we set $\lambda_B = 0.5$. For higher values of λ_B, U-turn events might be counted as reversals, that is, events where the magnetization precesses around $\mathbf{M}_B$ with a small angle, but ends up returning to A, in which case the reversal rate gets over-estimated. The attempt frequency f_0 is linked to the average waiting time τ_{AB} via

$$f_0 = \tau_{AB}^{-1} e^{\beta \Delta E}. \tag{5.15}$$

The results of Langevin simulations are shown in Fig. 5.2c as yellow triangles and agree well with Langer's theory.

5.3 The Forward Flux Sampling Method

Forward flux sampling (FFS) is a path sampling method for the simulation of rare events in systems with stochastic dynamics. It was initially developed by Allen et al. [9, 10] to simulate rare switching events in biochemical networks. It has, since then, been applied to a wide range of rare event problems [11]. In particular, it was used by Vogler et al. [8] to obtain switching rates in magnetic microstructures. In that context, FFS was shown to be significantly more efficient than direct Langevin simulations, while enabling an exploration of phase space free from prior assumptions. The method generates trajectories between two (meta)stable states A and B in a ratchet-like manner, which makes it well adapted for the simulation of rare to very rare events. Compared to other path sampling methods, it does not require prior knowledge of the phase space probability distribution function, which makes it suitable for both equilibrium and non-equilibrium systems—that is, systems that do not obey detailed balance. For equilibrium systems, detailed balance entails that the stationary phase space probability distribution is given by the Boltzmann distribution (Eq. (3.21)), and the dynamics is time-reversible. In non-equilibrium systems, on the other hand, the probability distribution function is not known a priori. Lastly, FFS requires the rate constant k_{AB} to be time-invariant, which is verified if the transition occurs between two time-invariant steady states A and B, and is itself fast compared to k_{AB}^{-1}. Following the initial work of Allen et al. [9, 10], many variants of FFS were developed [11]. In this work, we stick to the original algorithm, known as Direct FFS (DFFS), which we simply refer to as FFS.

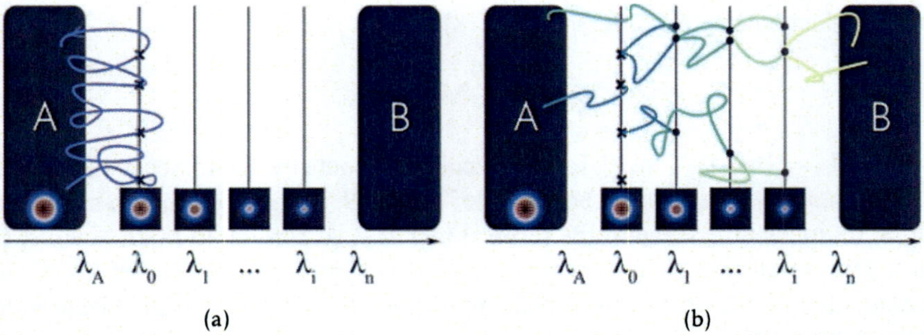

Fig. 5.3 Illustration of the FFS method where interfaces correspond to the isotropic collapse of a skyrmion. **a** During the first simulation starting from state A, trajectories (in purple) that exit region A and cross λ_0 are stored (marked as X's). **b** Langevin trial runs are performed at subsequent interfaces. The trajectories that successfully cross the next interface before returning to A are stored (marked as dots) and serve as starting points for the firing of new trial runs at the next interface. For the skyrmion collapse, we chose the order parameter defining the interfaces between states A (metastable skyrmion) and B (ferromagnetic state) as the monotonically decreasing size of the skyrmion

5.3.1 Principle

The FFS method employs a set of n (+1) non-intersecting interfaces in phase space to sample the transition path ensemble and compute a transition rate. The interfaces $\{\lambda_A, \lambda_0, \ldots \lambda_n = \lambda_B\}$ are defined as isosurfaces of a monotonically varying order parameter, $\lambda(x) = \lambda_i$, such that $\lambda_A < \lambda_0 < \ldots < \lambda_B$, as illustrated in Fig. 5.3. State A is defined as $\lambda(x) < \lambda_A$, and state B as $\lambda(x) > \lambda_B$. The other interfaces $\{\lambda_i\}$, $i = 0 \ldots n-1$ must satisfy $\lambda_i < \lambda_{i+1}$ $\forall i$. Any trajectory going from A to B must cross all the interfaces at least once. The boundaries of regions A and B, i.e., λ_A and λ_B, require appropriate definitions to ensure that the final rate is independent of the way they are defined, and only the efficiency of the method is affected.

The rate of transition from A to B within the FFS method is expressed as [9]

$$k_{AB} = \Phi_{A,n} = \Phi_{A,0} P(\lambda_n|\lambda_0). \tag{5.16}$$

Equation (5.16) translates the fact that $\Phi_{A,n}$, which is the flux of trajectories leaving state A and reaching interface $\lambda_n = \lambda_B$, is given by the product of $\Phi_{A,0}$—the flux of trajectories leaving A and crossing the first interface λ_0, and the probability $P(\lambda_n|\lambda_0)$ that a trajectory that crosses λ_0 coming from A will subsequently cross λ_n before returning to A. The flux $\Phi_{A,0}$ can be computed easily, since trajectories coming from A will frequently cross λ_0. On the other hand, the probability $P_B = P(\lambda_n|\lambda_0)$ will typically be small for rare events. It is made easier to calculate by decomposing it into a product of flux through each interface, that is

$$P_B = P(\lambda_n|\lambda_0) = \prod_{i=0}^{n-1} P(\lambda_{i+1}|\lambda_i). \tag{5.17}$$

The conditional probabilities $P(\lambda_{i+1}|\lambda_i)$ correspond to the probability that a trajectory coming from A that crossed interface λ_i for the first time will cross interface λ_{i+1} before returning to A. Equation (5.17) does not involve a Markovian approximation, since $P(\lambda_{i+1}|\lambda_i)$ depends on the history of trajectories that reached λ_i. Equations (5.16) and (5.17) correspond to an "effective positive flux" formulation of the rate constant. One can note that state B is not required to be a (meta)stable state, and FFS may be used to predict the probability of arbitrary rare fluctuations starting from a stable state A [12]. The terms in Eq. (5.16) are computed by firing Langevin trial runs at successive interfaces. The protocol is illustrated in Fig. 5.3 and detailed below. First, a single Langevin simulation is started in state A. This initial simulation is used to generate a collection of N_0 configurations corresponding to crossings of λ_0, as well as to compute a value of $\Phi_{A,0}$. Following an equilibration period, each time the system successfully exits region A and crosses the next interface λ_0, its configuration is stored. This is illustrated in Fig. 5.3a, where the system trajectory is shown in purple, and stored configurations are marked as X's. The simulation ends after N_0 crossing configurations have been stored at λ_0, and the flux of trajectories out of state A in Eq. (5.16) is obtained as $\Phi_{A,0} = N_0/\Delta t_{\text{sim}}$, in which Δt_{sim} is the total simulated time. Note that Δt_{sim} should not include the time the system may have spent in the B-basin.

After that, a configuration stored at λ_0 is selected at random, and used as a starting point for a new simulation. That new simulation ends when the trajectory either crosses interface λ_1, in which case the crossing configuration is stored, or the system returns to A. This is shown in Fig. 5.3b, where trajectories starting from λ_0 are shown in blue. If they cross the next interface, the stored configuration is marked as a dot. To obtain a flux through interface λ_1, this procedure is repeated M_0 times. If N_0^s trajectories successfully crossed λ_1, we can compute $P(\lambda_1|\lambda_0) = N_0^s/M_0$, where M_0 corresponds to the number of trial runs started at interface λ_0. One then proceeds analogously at subsequent interfaces, until λ_n is reached. This results in the system being driven in a ratchet-like manner from the initial to the final state, without imposing any bias on the microscopic dynamics. This is achieved by "capitalizing" on fluctuations in the direction of the order parameter. The FFS run terminates successfully if at least one trajectory reached state B, otherwise the interface definitions need to be improved. A collection of transition paths may be reconstructed *a posteriori* from the connectivity history of all the trial runs. Note that the obtained transition paths are branched, since many paths may start from a single configuration at λ_0.

5.3.2 Error Estimation and Optimization

5.3.2.1 Error Estimation

The FFS method involves many adjustable parameters: the number and position of the interfaces, $\{\lambda_i\}$, the definition of the order parameter, x, the number of trial runs fired at each interface, $\{M_i\}$, and the number of configurations stored at the first interface, N_0. The computed rate constant should, in principle, not depend on these parameters, and only affect the efficiency of the method [11]. In order to estimate the efficiency and the error of a given FFS run, a relative variance of the rate, $\mathcal{V}$, was derived in Ref. [13]. $\mathcal{V}$ is defined as the variance $V[k_{AB}]$ in the estimated value of the rate constant, k_{AB}, per starting point at λ_0, divided by the square of the expectation value of the rate:

$$\mathcal{V} = N_0 \frac{V[k_{AB}]}{k_{AB}^2} \approx N_0 \frac{\Phi_{A,0}^2 V[P_B]}{(\Phi_{A,0} P_B)^2} = N_0 \frac{V[P_B]}{P_B^2}. \tag{5.18}$$

In the previous expression, we have used the general formula

$$V[ax] = a^2 V[x], \tag{5.19}$$

where a is constant. In what follows, let $p_i \equiv P(\lambda_{i+1}|\lambda_i)$, $q_i \equiv 1 - p_i$, and $k_i \equiv M_i/N_0$. To obtain an expression for the variance of P_B, we use the following propagation formula:

$$V[f\{x_i\}] = \sum_{i=1}^{n} \left(\frac{\partial f}{\partial x_i}\right)^2 V[x_i], \tag{5.20}$$

where $f\{x_i\}$ is a function of n uncorrelated variables $(x_1, x_2, \ldots x_n)$, and the partial derivatives are evaluated with all variables at their mean values. We identify

$$f\{p_i\} = \prod_{i=0}^{n-1} p_i = P_B, \tag{5.21}$$

and, by noting that

$$\frac{\partial \prod_{j=0}^{n-1} p_j}{\partial p_i} = \frac{P_B}{p_i}, \tag{5.22}$$

we obtain

$$V[P_B] = V\left[\prod_{i=0}^{n-1} p_i\right] = P_B^2 \sum_{i=0}^{n-1} \frac{V[p_i]}{p_i^2}. \tag{5.23}$$

If we launch M_i trial runs at interface i, and N_i^s are successful in reaching the next interface, then, by using Eq. (5.19) once more,

$$V[p_i] = V\left[\frac{N_i^s}{M_i}\right] = \frac{V[N_i^s]}{M_i^2}. \tag{5.24}$$

According to [13], N_i^s is binomially distributed, and so $V[N_i^s] = M_i p_i q_i$. By inserting the previous expression and Eq. (5.24) into (5.23), we obtain

$$V[P_B] = \frac{P_B^2}{N_0} \sum_{i=0}^{n-1} \frac{q_i}{p_i k_i}, \tag{5.25}$$

and from Eq. (5.18),

$$\mathcal{V} = \sum_{i=0}^{n-1} \frac{q_i}{p_i k_i}. \tag{5.26}$$

Equation (5.26) holds as long as M_i is large enough that there is always at least one trial run that reaches the next interface. $\mathcal{V}$ is assumed to arise only from the trial runs for the computation of P_B (Eq. (5.17)), and not from the computation of the initial flux $\Phi_{A,0}$. This is justified for a large enough flux out of region A and a long enough simulation time in the A basin. In practice, we must chose N_0 so that the value of $\Phi_{A,0}$, and subsequent value of k_{AB}, do not strongly depend on it. Finally, we note that Eq. (5.26) was obtained by assuming that the number of successful trial runs at different interfaces are uncorrelated. In other words, the probability p_i of a successful trial run at λ_i does not affect the probability p_{i+1} of a successful trial run at λ_{i+1}. In reality, this assumption may not hold for highly correlated systems, and/or if the spacing between interfaces is small.

Statistical errors on the obtained transition rates can be extracted from the relative variance (Eq. (5.26)), or by repeating several FFS runs. In the former case, the error on k_{AB} may be estimated as [11, 13],

$$\sigma_k = k_{AB}\sqrt{\frac{\mathcal{V}}{N_0}}. \tag{5.27}$$

The statistical error on the attempt frequency f_0 is easily derived from Eq. (5.27), and we obtain

$$\sigma_{f_0} = f_0\sqrt{\frac{\mathcal{V}}{N_0}}. \tag{5.28}$$

Note that [8] shows that the standard deviation of the attempt frequency obtained by repeating five separate FFS simulations tends to be much bigger than the one computed with Eq. (5.28), as least in the case where $N_0 = 100$ was used. Apart from estimating a standard deviation, the relative variance is also useful to evaluate the computational efficiency of the method, and compare it with that of "brute force" direct Langevin simulations [13].

5.3.2.2 Optimization

Some iterative procedures were designed by Borrero and Escobedo [14] to optimize the parameters of FFS. The principle consists in minimizing the variance $V[P_B]$ under the constraint that the total transition probability P_B remains constant. Varying Eq. (5.25) with a Lagrange multiplier yields the condition [14]

$$M_i p_i = \text{cst}, \tag{5.29}$$

that is to say, a constant flux of partial trajectories through all interfaces is the optimal configuration to minimize the variance of the rate. From there, two main approaches are possible:

- optimizing the number of trial runs $\{M_i\}$ while keeping the interfaces fixed;
- optimizing the interface placement $\{\lambda_i\}$ at fixed $\{M_i\}$.

In this work, we have maintained a constant number of trial runs per interface, $M_i = M_0\ \forall i$, but, when judicious, we optimized the placement of interfaces. The principle is explained in what follows.

After choosing a fixed value of M_i through all interfaces, the condition (5.29) implies that

$$p_i = \text{cst} = (P_B)^{\frac{1}{n}}. \tag{5.30}$$

The value of the transition probabilities $\{p_i\}$ can be adjusted by changing the placement of the interfaces $\{\lambda_i\}$. To do so, an interpolation function is needed in order to establish a correspondence between the $\{p_i\}$ and the $\{\lambda_i\}$. A possible choice of such a function is [14]

$$f(\lambda_i) = \frac{\sum_{j=0}^{i-1} \ln p_j}{\sum_{j=0}^{n-1} \ln p_j}, \tag{5.31}$$

which reduces to

$$f(\lambda_i) = \frac{i}{n} \tag{5.32}$$

when $\{p_i\}$ = cst. The optimization procedure is as follows:

- perform a run of FFS with a non-optimized set of interfaces $\{\lambda_i\}$ and compute $\{f(\lambda_i)\}$ from Eq. (5.31);
- compute the optimum values of $\{f(\lambda_i)\}$ according to Eq. (5.32);
- invert the interpolation function to determine the new set of optimized $\{\lambda_i\}$;
- iterate the procedure if necessary.

5.3.3 Numerical Implementation

5.3.3.1 Implementation of Parallel Trial Runs

Although a linear implementation of FFS is straightforward, the method is more efficient when many trial runs at a given interface λ_i can be performed simultaneously. This requires the simulation to wait for all trial runs started at interface λ_i to terminate, in order to launch the next series of trial runs at interface λ_{i+1}.

All the code was written by the author, unless stated otherwise. The core code was implemented in C++ and performs a given trial run k, where $k = 1 \dots M_i$, at interface λ_i, until the system trajectory either returns to A or crosses λ_{i+1}. In the latter case, the configuration is stored in a file. The next layer is managed by a Bash script. The script stores the indices and the number N_i^s of crossing configurations saved from the trial runs fired at interface λ_{i-1}, and selects one at random to launch a trial run at interface λ_i. If $\lambda_i = \lambda_0$, it simply selects a random configuration in $[1, N_0]$. The FFS simulations were performed across several nodes—typically 10 nodes, with 16 tasks per node and two tasks per core, for a total of 320 simultaneous tasks—on the Vienna Scientific Cluster (VSC). Running parallel trial runs on multiple nodes requires the use of MPI (Message Parsing Interface) to allow the trial runs to wait for each other before moving on to the next interface. An MPI for Python script initially written by Dr. Christoph Vogler was modified by the author and used in that effect. The Python script calls the Bash script with some given parameters and manages the global FFS run over all interfaces and all trial runs on the VSC.

5.3.3.2 Numerical Test

Once more, we test the implementation against the rate of coherent magnetization reversal in a perpendicular field (Fig. 3.6a). We reproduce the low damping case of Ref. [8], for which we use the same parameters describing a single-macrospin particle with $\alpha = 0.002$, $\mathscr{K} = 3$ MJ/m^3, $a = 1$ nm, $N = 1$, $\mathscr{M}_s = 0.5\mu_0^{-1}$ A/m. Region A and B are defined according to Eq. (5.14), with $\lambda_{A,B} = 0.45$. The other interfaces are defined as hyperplanes in the space of configurations in the following way. We use the collection of images along the MEP, $[\mathbf{M}^1 \dots \mathbf{M}^Q]$, previously found with the GNEB scheme. Note that we may skip the images that lie past the saddle point, since the flux of trajectories starting from the saddle point that subsequently reach state B is high. Therefore, if the climbing image is at the saddle point, we can restrict ourselves to $[\mathbf{M}^1 \dots \mathbf{M}^{\nu_{CI}}, \mathbf{M}^Q]$, where $\mathbf{M}^Q = \mathbf{M}^B$. At each spin i, and each image ν, we define the local vector to the previous configuration,

$$\mathbf{n}_i^\nu = \mathbf{m}_i^\nu - \mathbf{m}_i^{\nu-1}. \tag{5.33}$$

Interface ν is defined as the hyperplane normal to $\mathbf{N}^\nu = [\mathbf{n}_1^\nu \dots \mathbf{n}_N^\nu]$ that passes through $\mathbf{M}^\nu$. At each time step, we compute the dot product

$$\mathbf{N}^{\nu} \cdot [\mathbf{M}(t) - \mathbf{M}^{\nu}]. \tag{5.34}$$

We consider that interface ν has been crossed if the dot product in Eq. (5.3) has changed its sign. Note that, for the problem of magnetization reversal, one may also use Euclidian norm definitions (Eq. (5.14)) for all interfaces [15]. We use eight interfaces, $N_0 = 100$, and $M_i = M_0$ =3840 at all interfaces. The temperature in the Langevin trial runs is chosen such that $\beta \Delta E = 14$. For a reduced field $h = 0.1$ (Eq. (3.93)), we obtain an initial flux of $\Phi_{A,0} = 1.06$ GHz, and an attempt frequency of f_0 = 147.5 GHz, with a relative variance $\mathcal{V} = 0.48$. This matches the result from Ref. [8] and the analytical formula derived in [16].

5.4 Application to Skyrmion Collapse and Comparison with Langer's Theory

5.4.1 Interface Definition

We aim to compare the collapse rates of an isolated skyrmion previously calculated via Langer's theory, with the rates computed via FFS simulations. A natural choice of interfaces for the skyrmion collapse is the decreasing size of the skyrmion, where state A is the equilibrium skyrmion size, and state B—corresponding to the ferromagnetic state—is associated to a zero-size. Arbitrarily, we consider that magnetic sites $\mathbf{m}_i$ that satisfy $m_{z,i} \leq 0$ are part of the skyrmion, and we define the order-parameter x as the (integer) number of sites inside the skyrmion,

$$x = \sum_{\substack{i=1 \\ \mathbf{m}_{z,i} \leq 0}}^{N} 1. \tag{5.35}$$

The interfaces $\lambda_i = \lambda(x)$ are isosurfaces of x (Fig. 5.3). We set $\lambda_A = \lambda(x_A)$, where x_A is the number of sites in the metastable skyrmion, and $\lambda_n = \lambda_B = \lambda(x = 0)$. From there, we may choose, for instance, the first interface to be $\lambda_0 = \lambda(x_A - 3)$. This ensures that the first simulation out of the A basin will span over a long enough simulated time. The rest of the interfaces may be evenly spaced out between λ_0 and λ_B. However, the efficiency of FFS is increased if the flux of partial trajectories remains as constant as possible through all interfaces. Therefore, we usually saturate the interfaces (one interface every integer value of x) in the region preceding the saddle point, since the flux typically tends to drop in that region. The rest of the interfaces can be more spaced out. After a first run of FFS has terminated, it can be helpful to run the interface optimization procedure described in Sect. 5.3.2.1. However, this scheme is designed for a continuous order parameter, and is not as effective if the order parameter only takes integer values. Although we did not implement it, a good solution in this case would be to also optimize the number of trial runs per interface

Table 5.2 For the FFS runs at $\beta\Delta E_1 = 10$ and for each value of the reduced field b_z, we give: the number of interfaces, the absolute temperature, the number of configurations stored at the first interface, N_0, the number of trial runs per interface, M_0, and the relative variance of the rate, $\mathcal{V}$

b_z (J_{ex})	0.000	0.005	0.010	0.015	0.020	0.025
Number of interfaces	34	32	24	22	20	17
$\Delta E_1/10k_B$ (K)	327.5	283.2	245.1	210.0	182.1	156.1
N_0	1500	1000	1000	1000	1000	1000
M_0	34560	27520	37760	38080	39680	25920
$\mathcal{V}$	1.01	0.86	0.76	0.70	0.58	1.00
b_z (J_{ex})	0.030	0.035	0.040	0.045	0.050	
Number of interfaces	17	17	14	14	14	
$\Delta E_1/10k_B$ (K)	132.8	111.8	92.9	76.2	61.2	
N_0	1000	1000	1000	1000	1000	
M_0	39680	39680	39680	39680	39680	
$\mathcal{V}$	0.66	0.73	0.73	0.84	0.85	

$\{M_i\}$ while maintaining a constant set of $\{\lambda_i\}$. Lastly, for higher field values, at which the metastable skyrmion is small, we simply saturate the interfaces between λ_0 and λ_B.

The number of stored configurations at λ_0 was chosen to be $N_0 = 1000$, or 1500 for $b_z = 0$. Since the trajectories are branched, we have found that the exhaustiveness of this initial sampling is crucial. If not enough crossing configurations are sampled, the fluxes of trajectories through subsequent interfaces are also not accurate. As for the number of trial runs per interface, it is chosen constant and equal to M_0 at all interfaces, and is iteratively increased until the variance (Eq. (5.26)) verifies $\mathcal{V} \lessapprox 1$. In some cases, the variance was decreased also by increasing the number of interfaces (p_i increases). In the end, the number of interfaces spans from 34 at zero-field, to 14 for $b_z = 0.05$. For each value of the reduced field at $\beta\Delta E_1 = 10$, Table 5.2 shows the corresponding number of interfaces, as well as the values of the absolute temperature of the trial runs, the number of trial runs performed at each interface, M_0, the number of configurations stored at the first interface, N_0, and the relative variance of the rate, $\mathcal{V}$.

5.4.2 Results

For each value of the reduced field, we compute a total collapse rate k_{AB} through forward flux sampling, and an attempt frequency is recovered by using the value

Table 5.3 For $b_z = 0.010$ and $\beta\Delta E_1 = 10$, we give the interface index, i, the number of sites inside the skyrmion, x, and the flux $p_{i-1} = P(\lambda_i|\lambda_{i-1})$ through λ_i

i	1	2	3	4	5	6	7	8	9	10	11	12
x	26	23	20	19	18	17	16	15	14	13	12	11
p_{i-1}	0.18	0.27	0.35	0.70	0.68	0.67	0.65	0.63	0.59	0.57	0.54	0.52
i	13	14	15	16	17	18	19	20	21	22	23	
x	10	9	8	7	6	5	4	3	2	1	0	
p_{i-1}	0.49	0.48	0.43	0.39	0.39	0.37	0.40	0.28	0.53	0.77	0.99	

of $\Delta\bar{E}$ (Eq. (4.3)). Throughout each FFS run, we maintain a temperature T such that $\beta\Delta E_1 = 10$ or 15, so that Langer's theory holds [17]. For $b_z = 0.010$, we get $\Phi_{A,0} = 104.39$ GHz with $\lambda_0 = \lambda(x = 29)$. For this particular field value, FFS matches really well with Langer's theory, and so we give in Table 5.3 the values of the order parameter x, and the corresponding flux p_{i-1} through each interface λ_i, in the case where $\beta\Delta E_1 = 10$. The results from both FFS and Langer's theory are shown in Fig. 5.4a for the attempt frequency f_0, and the skyrmion lifetime in an infinite system, $\tau_{AB} = k_{AB}^{-1}$, is plotted in Fig. 5.4b. The standard deviation on the lifetime is derived from Eq. (5.27) and is expressed as

$$\sigma_\tau = \frac{1}{k_{AB}}\sqrt{\frac{\mathcal{V}}{N_0}}. \tag{5.36}$$

For the attempt frequency, the formula is modified to account for the error on the internal energy barrier in the following way:

$$\sigma_{f_0} = f_0\Big(\sqrt{\frac{\mathcal{V}}{N_0}} + \beta\sigma_{\Delta E}\Big), \tag{5.37}$$

where $\sigma_{\Delta E}$ is the standard deviation associated with $\Delta\bar{E}$. At higher applied field values, where the attempt frequency is found within the 1–10 GHz range, we also compute a collapse rate via direct Langevin simulations at 300 K. In that case, we proceed similarly to the FFS method, by counting the number of sites x inside the skyrmion at each time step, as defined in Eq. (5.35). A collapse is recorded when $x = 0$. We obtain an average lifetime out of 400 collapses, and once again we use the value of $\Delta\bar{E}$ to recover an attempt frequency.

Overall, we obtain a good agreement between Langer's predictions and FFS. The three data points obtained from direct Langevin simulations match the result of FFS. The prefactors obtained at two different temperatures in the FFS runs agree well, which validates Langer's result that, in the present case, the attempt frequency does not exhibit any significant temperature dependence. Over the range of applied field

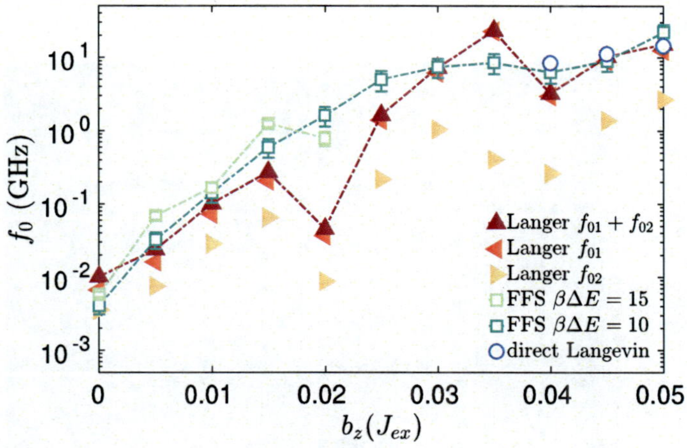

(a)

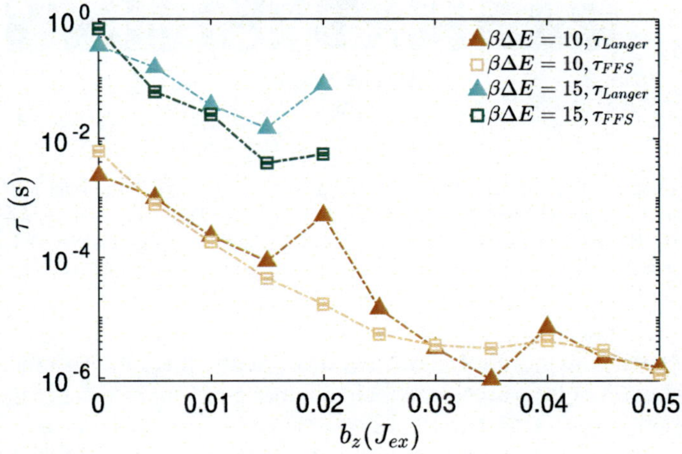

(b)

Fig. 5.4 We show the following quantities as a function of the reduced destabilizing applied field b_z: **a** Attempt frequency f_0 for the skyrmion collapse obtained via Langer's theory calculations, FFS simulations, and direct Langevin simulations at 300 K, **b** skyrmion lifetime against collapse, $\tau = k_{AB}^{-1}$, obtained via Langer's theory and FFS simulations. The results from Langer's theory are the ones presented in Chap. 4. The errorbars on FFS calculations are respectively given by Eqs. (a) (5.37) for the attempt frequency and (b) (5.27) for the average lifetime. Note that whenever the errorbars are smaller than the size of the square marker, they are represented as two unconnected parallel lines

strengths studied here, FFS confirms the prediction that the prefactor f_0 varies by four orders of magnitude, which is the result of a strong entropic contribution.

In Fig. 5.5, we plot some examples of stored configurations at interfaces at which the order-parameter is that of a saddle point. We report both S_1 and S_2 types of configurations, as well as some other configurations that don't clearly pertain to

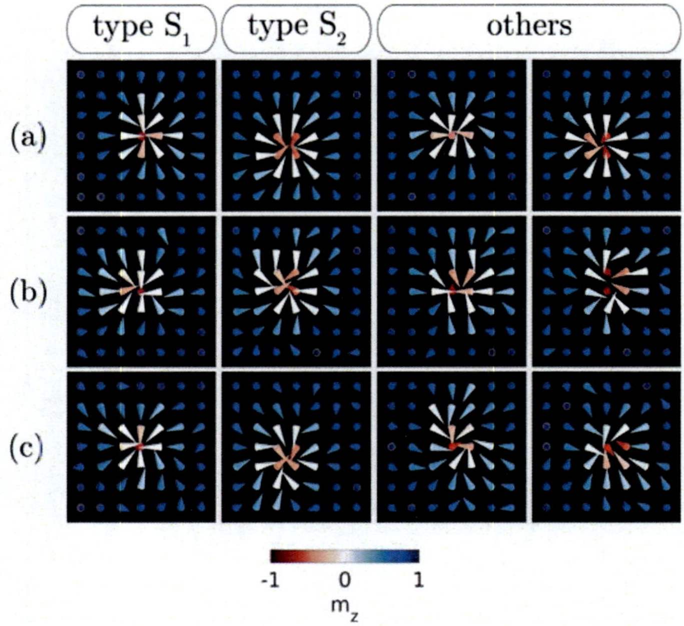

Fig. 5.5 Examples of (zoomed) spin configurations from FFS stored at the interface(s) at which the order-parameter equates that of a saddle point **a** at interface $\lambda_9 = \lambda(x = 4)$, for $b_z = 0.05$ and $T = 61\,\text{K}$ ($\beta\Delta E_1 = 10$), **b**, **c** for $b_z = 0$ and $T = 218\,\text{K}$ ($\beta\Delta E_1 = 15$), at interface **b** $\lambda_{28} = \lambda(x = 5)$, and **c** $\lambda_{29} = \lambda(x = 4)$

either category. This occurs for two reasons. Firstly, the crossing configurations correspond to an order parameter which is either equal to, or smaller than that of the saddle point ($x \leq x_{\text{SP}}$). Even in cases where $x = x_{\text{SP}}$, this does not imply that the configuration is indeed a saddle point. Secondly, the saddle point is the lowest energy configuration at the barrier top. Under the effect of thermal fluctuations, the system does not usually cross the barrier exactly at the SP, but deviates from it by a more or less small amount. Note that Langer's theory is in principle valid only for small deviations.

5.5 Discussion

In this chapter, we have compared the rate of collapse of an isolated skyrmion computed via Langer's theory and forward flux sampling simulations. Overall, both methods agree well, thus, in general, the choice of either method should yield the correct order of magnitude for the skyrmion lifetime. Therefore, we can confirm the result previously obtained from Langer's theory that the stability of skyrmions is subject to a strong entropic contribution, and that internal energy barriers are not enough to properly predict their lifetime.

For some values of the applied field (most notably, $b_z = 0.02$), the attempt frequency predicted by Langer shows a discrepancy of one or two orders of magnitude with the FFS result. In general, since the application of Langer's theory is bound by many hypotheses concerning the thermal fluctuations, the energy landscape, and the lack of coupling between different mechanisms, the result of FFS should be more trustworthy. On the other hand, FFS is less computationally efficient and less systematic in its application, because the interfaces need to be adjusted to reflect different sets of parameters. The computational cost of both Langer and FFS will obviously increase with the system size, and if non-local dipolar couplings are included, but the cost of FFS will in principle increase faster. For different types of processes, such as escape processes, the nature of the order parameter should be modified accordingly.

As far as we are aware, the results in this chapter constitute the first comparison between forward flux sampling and Langer's theory in a complex, nontrivial magnetic system that cannot easily be treated analytically (i.e., beyond the macrospin approximation). In particular, in addition to the path over the first-order saddle point, S_1, we included the contribution of a path involving a third-order saddle point, S_2, to Langer's result in Chap. 4. Multidimensional formulations of reaction rate theory are typically found for a transition path involving a first-order saddle point, and only Ref. [18] proposed an extension of Langer's theory to include higher-order saddle points. Nevertheless, the activation energies associated with S_2 are almost the same as the ones associated with S_1, and configurations similar to S_2 are observed in FFS, so we could not justify neglecting it a priori. Only through the computation of the attempt frequencies in Chap. 4 did it transpire that the mechanism involving S_2 should not contribute as much as the one involving S_1, because the attempt frequencies for mechanism n°2 tend to be much lower (see Fig. 5.4a). This is because the configurational entropy at S_2 is lower than at S_1, which implies that S_2 is less likely to be visited by the system. This result seems to be confirmed by FFS simulations, since the obtained attempt frequencies tends to match Langer's prediction. This is further proof of the importance of the prefactor, because, even though $\Delta E_1 \approx \Delta E_2, k_1(T) \gg k_2(T)$ (see Fig. 4.14d). We should also note that the energy barriers between S_1 and S_2 are likely to be quite low, therefore, in reality, both mechanisms are probably coupled in some way.

Finally, at zero-field and $T = 200$ K, the skyrmion's average lifetime against collapse approaches one second, which may be encouraging for applications. In multilayers, where the layers are stacked along the z-direction, the energy barriers should increase if the spin texture is homogeneous along z, thus enhancing stability [19]. Additionally, in Pd/Fe bilayers on Ir(111), frustrated exchange has also been shown to increase the height of the activation barriers against collapse [20]. In the next chapter, we will look at the paths to annihilation of skyrmions under frustrated exchange couplings.

References

1. García-Palacios JL, Lázaro FJ (1998) Langevin-dynamics study of the dynamical properties of small magnetic particles. Phys Rev B 58:14937–14958
2. Desplat L, Vogler C, Kim J-V, Stamps RL, Suess D (2020) Path sampling for lifetimes of metastable magnetic skyrmions and direct comparison with Kramers' method. Phys Rev B 101(6):060403(R)
3. Brown WF (1963) Thermal fluctuations of a single-domain particle. Phys Rev 130:1677–1686
4. Coffey W, Kalmykov YP (2012) The Langevin equation: with applications to stochastic problems in physics, chemistry and electrical engineering, vol 27. World Scientific
5. Risken H (1996) The Fokker-Planck equation. Springer series in synergetic, vol 18. Springer, Berlin, Heidelberg
6. Matsumoto M, Nishimura T (1998) Mersenne twister: a 623-dimensionally equidistributed uniform pseudo-random number generator. ACM Trans Model Comput Simul (TOMACS) 8(1):3–30
7. Scholz W, Schrefl T, Fidler J (2001) Micromagnetic simulation of thermally activated switching in fine particles. J Magn Magn Mater 233(3):296–304
8. Vogler C, Bruckner F, Bergmair B, Huber T, Suess D, Dellago C (2013) Simulating rare switching events of magnetic nanostructures with forward flux sampling. Phys Rev B 88(13):134409
9. Allen RJ, Warren PB, ten Wolde PR (2005) Sampling rare switching events in biochemical networks. Phys Rev Lett 94:018104
10. Allen RJ, Frenkel D, ten Wolde PR (2006) Simulating rare events in equilibrium or nonequilibrium stochastic systems. J Chem Phys 124(2):024102
11. Allen RJ, Valeriani C, ten Wolde PR (2009) Forward flux sampling for rare event simulations. J Phys Condens Matter 21(46):463102
12. Borrero EE, Escobedo FA (2009) Simulating the kinetics and thermodynamics of transitions via forward flux/umbrella sampling. J Phys Chem B 113(18):6434–6445
13. Allen RJ, Frenkel D, ten Wolde PR (2006) Forward flux sampling-type schemes for simulating rare events: efficiency analysis. J Chem Phys 124(19):194111
14. Borrero EE, Escobedo FA (2008) Optimizing the sampling and staging for simulations of rare events via forward flux sampling schemes. J Chem Phys 129(2):024115
15. Vogler C, Bruckner F, Suess D, Dellago C (2015) Calculating thermal stability and attempt frequency of advanced recording structures without free parameters. J Appl Phys 117(16):163907
16. Kalmykov YP (2004) The relaxation time of the magnetization of uniaxial single-domain ferromagnetic particles in the presence of a uniform magnetic field. J Appl Phys 96(2):1138–1145
17. Coffey W, Garanin D, McCarthy D (2001) Crossover formulas in the Kramers theory of thermally activated escape rates—application to spin systems. Adv Chem Phys 117:483–765
18. Duff G (2008) Langer's method for the calculation of escape rates and its application to systems of ferromagnets. PhD thesis, Dublin Institute of Technology
19. Moreau-Luchaire C, Moutafis C, Reyren N, Sampaio J, Vaz C, Van Horne N, Bouzehouane K, Garcia K, Deranlot C, Warnicke P et al (2016) Additive interfacial chiral interaction in multilayers for stabilization of small individual skyrmions at room temperature. Nat Nanotechnol 11(5):444–448
20. von Malottki S, Dupé B, Bessarab P, Delin A, Heinze S (2017) Enhanced skyrmion stability due to exchange frustration. Sci Rep 7(1):12299

Chapter 6
Paths to Annihilation of First- and Second-Order (Anti)skyrmions Under Frustrated Exchange

6.1 Introduction

As we have seen in Sect. 2.3.4, skyrmionic solutions can exist in centrosymmetric systems when, for instance, a characteristic lengthscale is introduced via exchange frustration. Systems with frustrated exchange are particularly interesting, as they allow many different kinds of topological defects to coexist. Frustrated exchange was shown to arise from interface effects in Pd/Fe/Ir(111) for fcc stackings of the Pd layer by DFT calculations. This system was also shown to host nanoscale interface skyrmions stabilized primarily by the DMI [1, 2]. The ab initio calculations in Ref. [2] were in good agreement with the different phases of the system reported from SP-STM experiments [1]. More recently, a strong exchange frustration was observed in Rh/Fe/Ir(111), which was found to be a driving force behind the SS ground state [3]. Lastly, during the writing phase of this thesis, strong exchange frustration was reported in Rh/Co/Ir(111) by SP-STM measurements and DFT calculations [4]. In this system, skyrmions of approximately 5 nm in diameter were observed at zero magnetic field at low temperature in the virgin FM state. The exchange frustration was found to play an important role in their stabilization. These results demonstrate the importance of understanding the effect of exchange frustration on skyrmion stability. The frustrated $J_1 - J_2$ system on the hexagonal lattice was extensively investigated in Ref. [5] and, in general, the frustrated hexagonal lattice is the most commonly studied [2, 6–9]. Although skyrmions decouple from the lattice in the long wavelength limit, we can note that this is typically not true for saddle point configurations of collapse processes [10, 11]. Collapse mechanisms of skyrmions and antiskyrmions in chiral thin films with frustrated exchange were previously investigated [7, 10], and appear similar to the isotropic-type of collapse calculated in non-frustrated chiral systems [11–16]. In Pd/Fe/Ir(111), the energy barriers of isolated skyrmions with respect to isotropic collapse were found to be enhanced by the exchange frustration, compared to a Hamiltonian limited to first nearest-neighbours [7]. A third annihilation mechanism involving the injection of a singularity with

L. Desplat, *Thermal Stability of Metastable Magnetic Skyrmions*, Springer Theses,
https://doi.org/10.1007/978-3-030-66026-0_6

opposite topological charge was reported in [15], but the path appears to go over a higher order saddle point. Additionally, this mechanism seems to emerge for larger skyrmion sizes. Lastly, asymmetric paths to collapse similar to the ones we present in this chapter were very recently reported via GNEB calculations for first-order skyrmion solutions on the frustrated hexagonal lattice [4, 17].

In this chapter, we look at paths to annihilation of first- and second-order skyrmions and antiskyrmions on the frustrated $J_1 - J_2 - J_3$ square lattice previously studied by Lin et al. [18], which we presented in Sect. 2.3.4. We relax skyrmionic solutions spanning over only a few sites in diameter that, once again, do not exhibit translational invariance on the lattice, and instead experience pinning at particular lattice points. By using the GNEB method, we compute a new type of asymmetric path to collapse. A weak interfacial DMI is added to break the invariance with respect to the rotation of helicity (i.e. the "spin" degree of freedom, where "spin" refers to the spin of the skyrmion as a particle [18]), as well as the degeneracy between skyrmions and antiskyrmions. The results presented in this chapter were initially published in Ref. [19].

The chapter is organized as follows. In Sect. 6.2, we give regions of existence of metastable first- and second-order skyrmions and antiskyrmions in parameter space. In Sect. 6.3, we present different mechanisms by which they annihilate, which we then validate by means of direct Langevin simulations in Sect. 6.4. We subsequently compute attempt frequencies associated with the different mechanisms by applying Langer's theory in Sect. 6.5, and the average lifetimes of the structures are discussed. Finally, the results are discussed in Sect. 6.6.

6.2 Coexistence of Skyrmion and Antiskyrmion Solutions

The system we study is similar to the one described in Chap. 4, but the isotropic exchange is now extended to the third shell of neighbours, with the following reduced coupling parameters: $(j_1, j_2, j_3) = (1, -0.35, -0.15)$ [18]. The reduced perpendicular uniaxial anisotropy constant is set to $k = 0.1$ for the rest of this chapter. We introduce a weak DMI to the system in order to break the invariance with respect to the rotation of helicity, as well as the degeneracy between skyrmions and antiskyrmions. We look for a set of parameters that allow metastable skyrmion and antiskyrmion solutions to co-exist. To do so, we initialize the system close to a(n) (anti)skyrmion state and run the iterative energy minimization procedure described in Sect. 3.5.1. If a(n) (anti)skyrmion solution exists, it is relaxed. We vary the perpendicular destabilizing applied field b in [0, 0.4], and the DMI coupling constant d in [0, 0.07]. We obtain the diagrams of Fig. 6.1 for skyrmions (Fig. 6.1a) and antiskyrmions (Fig. 6.1b).

At low field or high DMI, a spiral state is relaxed. Above a critical applied field, the spin-spiral state becomes the fully polarized ferromagnetic state via a second-order phase transition [18]. Within the FM phase, metastable skyrmion solutions can be stabilized. In the presence of a perpendicular easy-axis anisotropy, the skyrmion

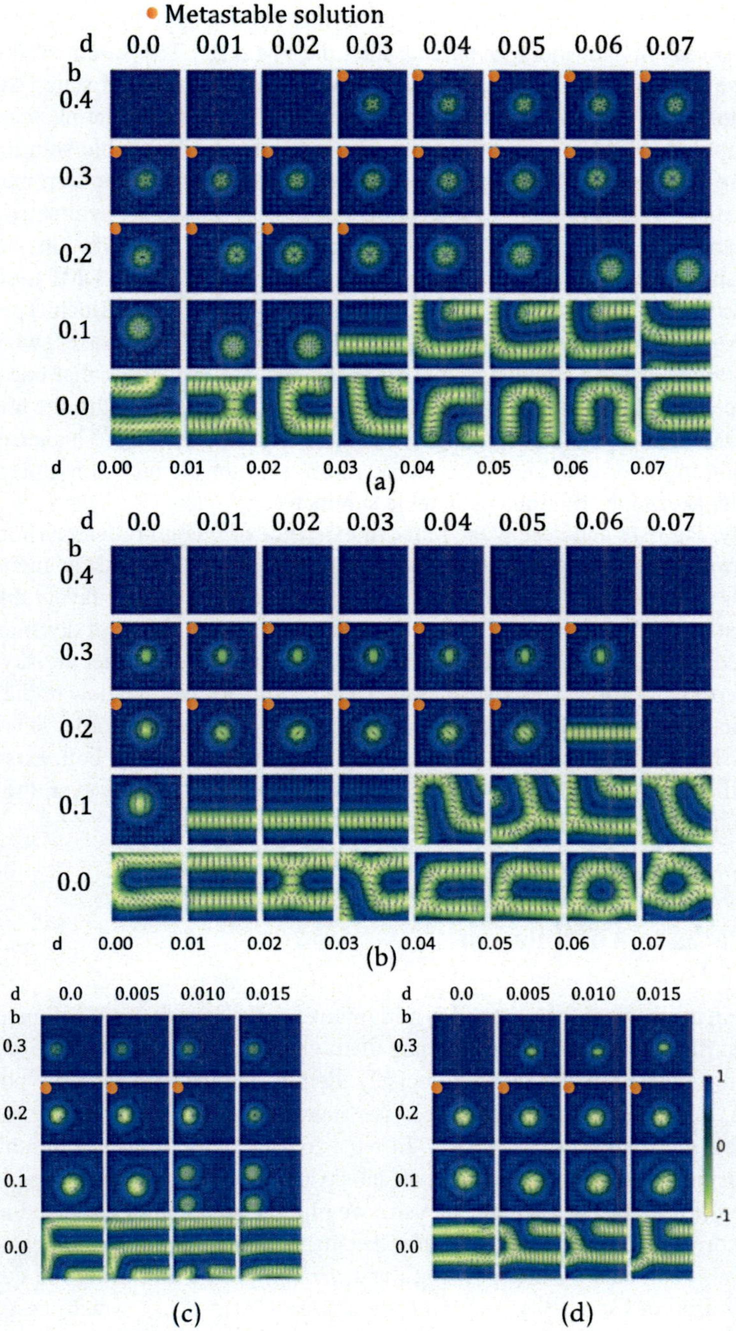

Fig. 6.1 For $k = 0.1$ and different values of the reduced DMI constant d and the reduced applied field b, existence of **a** an isolated skyrmion solution, **b** an isolated antiskyrmion solution, **c** an isolated second-order skyrmion solution, **d** an isolated second-order antiskyrmion solution. Only metastable solutions of interest corresponding to excitations of the ferromagnetic ground state are marked with an orange dot. The spin maps are zoomed in to show the topological defects

lattice becomes thermodynamically stable at intermediate applied field [18]. In this region, skyrmions are more favourable than the FM state. They are metastable in the sense that they are local energy solutions, but not the ground state. At finite temperature, the isolated skyrmion state is rapidly destroyed by the nucleation of many topological defects in its vicinity. By contrast, the metastable solutions we are interested in are isolated skyrmionic defects as excitations of the ferromagnetic ground state, which we mark as orange dots in Fig. 6.1. Since the symmetry of the DMI favours Néel skyrmions, degeneracies are lifted, and antiskyrmions become progressively less stable than their skyrmion counterparts as the DMI increases. They also relax diagonally on the square lattice, such that their Bloch-type axes, which we showed in Fig. 2.6, are along the first nearest-neighbour axes, which are the only ones that are coupled by the DMI. This is due to the fact that one of the Néel-type axes of the antiskyrmion exhibits the opposite chirality to the one favoured by the DMI, as was illustrated in Fig. 2.6 and discussed in Sect. 2.3. This introduces frustration in the orientation of the antiskyrmion solutions, which is not present in first-order skyrmion solutions with cubic symmetry.

Lastly, Fig. 6.1c and d show the range of existence of second-order skyrmion and antiskyrmion solutions (see Fig. 2.9d and e), respectively. They exist as metastable solutions at low d in the skyrmion lattice phase, where they are more favourable than the FM state. By increasing the DMI at constant applied field, a bound skyrmion pair is relaxed instead of the second-order skyrmion, but the second-order antiskyrmion solution persists. Metastable solutions are found within a small window of the ferromagnetic phase at sufficiently low field and DMI. We find second-order skyrmions and antiskyrmions to be quasi-degenerate in energy whenever they both exist, with small differences most likely caused by the underlying lattice. However, the range of existence of the antiskyrmion appears to be wider.

6.3 Paths to Annihilation

Transition mechanisms correspond to minimum energy paths on the $2N$-dimensional energy surface. Once more, we compute them via CI-GNEB [12] calculations with a climbing image scheme [20] to precisely identify the first-order saddle point on the path. In the GNEB method, successive states of the system along the reaction coordinate are referred to as images. In what follows, we present the annihilation mechanisms of a skyrmion (Fig. 6.2), an antiskyrmion (Fig. 6.4), and a second-order skyrmion (Fig. 6.6). For each mechanism, we plot the spin maps at selected images and the corresponding topological charge density, ρ_s. Note that the topological charge density was obtained by interpolating the skyrmion profile and using the continuous definition of Eq. (2.34). However, the discrete form of ρ_s, which we gave in Eq. (2.39), is the more correct definition that should be used on the atomistic spin lattice. The total energy along the path is plotted in Figs. 6.3, 6.5, and 6.7 for the skyrmion, the antiskyrmion, and the second-order skyrmion, respectively, in units of the isotropic exchange coupling constant between first neighbours, J_1, as well as the total topological charge, N_s, at each image, which was calculated according to the discrete definition in Eq. (2.38).

6.3.1 Skyrmion

The first set of parameters we examine is $(b, d) = (0.2, 0.03)$. It is a region of parameter space where both metastable skyrmion and antiskyrmion solutions exist. In this context, we observe an annihilation mechanism which is different from the usually reported isotropic type of collapse. We show spin maps at selected images along the path in Fig. 6.2a, and the corresponding energy profile is shown in red in Fig. 6.3a, where the evolution of the topological charge is shown in Fig. 6.3b. The mechanism is as follows. One half of the skyrmion unfolds, and a half-antiskyrmion, or antimeron, is nucleated in its place (ims. [1-6]). The injection of the opposite topological charge in the system constitutes the first-order saddle point on the transition path (im. 6 in Fig. 6.2a and inset on the right in Fig. 6.3a), and, as shown in Fig. 6.3b, also coincides with the loss of the topological charge. This state possesses four different realizations, equivalent to rigid $\pi/2$-rotations of the whole system. The remaining meron and antimeron then naturally annihilate (ims. [7-10]). We found that the isotropic type of collapse also exists in this system, but it involves a higher order saddle point.

The second set of parameters is $(b, d) = (0.3, 0.07)$. In the space of parameters, it is situated just above the limit at which antiskyrmion solutions no longer exist. We show spin maps at selected images along the path in Fig. 6.2b, and the corresponding energy profile is shown in pale blue in Fig. 6.3a, where the evolution of the topological charge is shown in Fig. 6.3b. In this case, the skyrmion undergoes a regular collapse involving a first-order saddle point (im. 6 in Fig. 6.2b and inset in the centre in Fig. 6.3a), but the core at the SP displays a weak asymmetry in a way that is reminiscent of the mechanism in Fig. 6.2a. For this reason, the saddle point is different from the saddle point configurations S_1 and S_2 that we reported in Chap. 4 in the case of the isotropic collapse of a chiral skyrmion. Since we are close to the region of antiskyrmions' metastability, this mechanism can be considered as intermediate between antimeron nucleation and isotropic collapse.

Finally, we select a region of parameter space with high DMI, far from the metastability region of antiskyrmions: $(b, d) = (0.7, 0.2)$. We show spin maps at selected images along the path in Fig. 6.2c, and the corresponding energy profile is shown in dark blue in Fig. 6.3a, where the evolution of the topological charge is shown in Fig. 6.3b. As expected, we obtain a perfectly isotropic collapse similar to the ones reported in Chap. 4, and without the injection of an opposite topological charge. In all three cases, the topological charge seems to fall to zero at the saddle point, although a higher resolution in the SP region could potentially reveal a state with a half-integer charge preceding the SP.

Similar findings were reported very recently in Ref. [4]. In this work, the authors observed 5-nm skyrmions at zero magnetic field in the virgin FM phase of Rh/Co on Ir(111) via spin polarized STM. DFT calculations revealed a strong exchange frus-

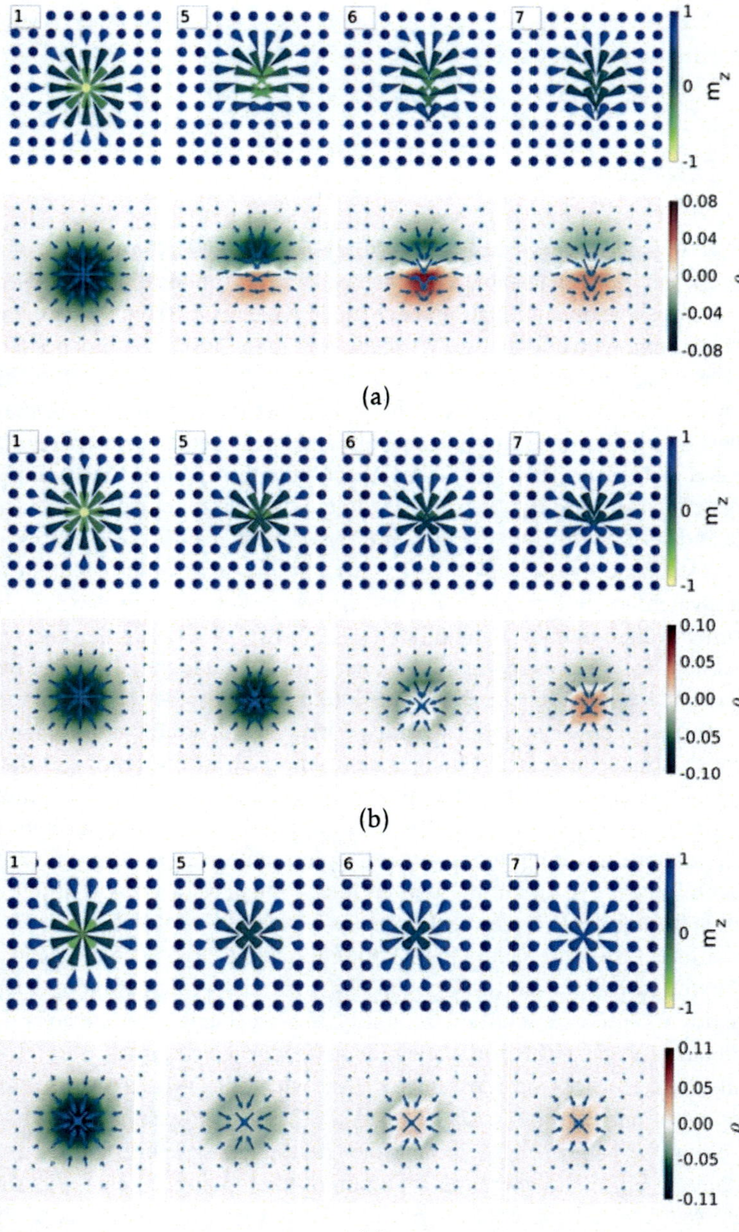

Fig. 6.2 Spin maps (zoomed) and corresponding topological charge density along the transition path for a skyrmion annihilation. The parameters are **a** $(b, d) = (0.2, 0.03)$ (metastable antiskyrmion solutions exist), **b** $(b, d) = (0.3, 0.07)$ (close to the existence of antiskyrmion solutions), **c** $(b, d) = (0.7, 0.2)$ (antiskyrmions solutions do not exist). The image index is given in the top left-hand corner. When antiskyrmion solutions exist, the skyrmion annihilates via the nucleation of a half-antiskyrmion, or antimeron. The meron-antimeron state constitutes the saddle point (im. 6). When antiskyrmion solutions do not exist, the isotropic collapse is recovered

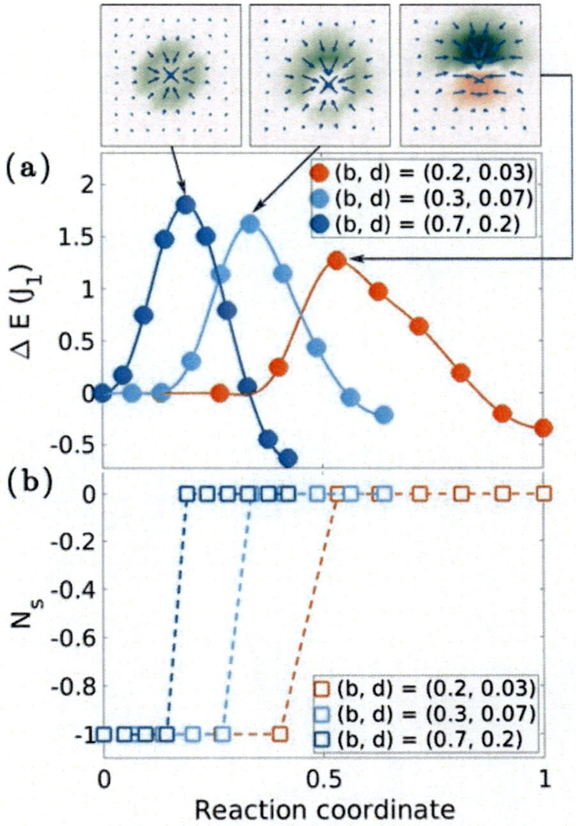

Fig. 6.3 For the skyrmion annihilation at different values of the reduced applied field b and reduced DMI constant d, we show the evolution along the transition path of **a** the internal energy barrier in units of J_1, **b** the topological charge. The reaction coordinate is normalized by the largest path length. The insets show the spin configuration and the topological charge density at the saddle point. In all three cases, the SP has a vanishing topological charge

tration in this system. Subsequent atomistic GNEB simulations with a Hamiltonian parametrized by DFT showed that at zero field, the preferred collapse mechanism is similar to the one we reported here at low field and DMI in Fig. 6.2a, where the barrier results mainly from the strong isotropic exchange coupling between first nearest-neighbours, J_1, more than from the DMI. At a higher applied field of 1T, the isotropic collapse was recovered, in agreement with the mechanisms that we presented in Fig. 6.2b and c at higher field and DMI. In that case, isotropic collapse was preferred as it lowers the anisotropy and the Zeeman energy contributions. The existence of antiskyrmions in the system was not reported.

6.3.2 Antiskyrmion

We consider again the first set of parameters, $(b, d) = (0.2, 0.03)$, which allows metastable antiskyrmions to coexist with skyrmions. We look for paths to annihilation of an antiskyrmion, and we relax two different mechanisms passing through a first-

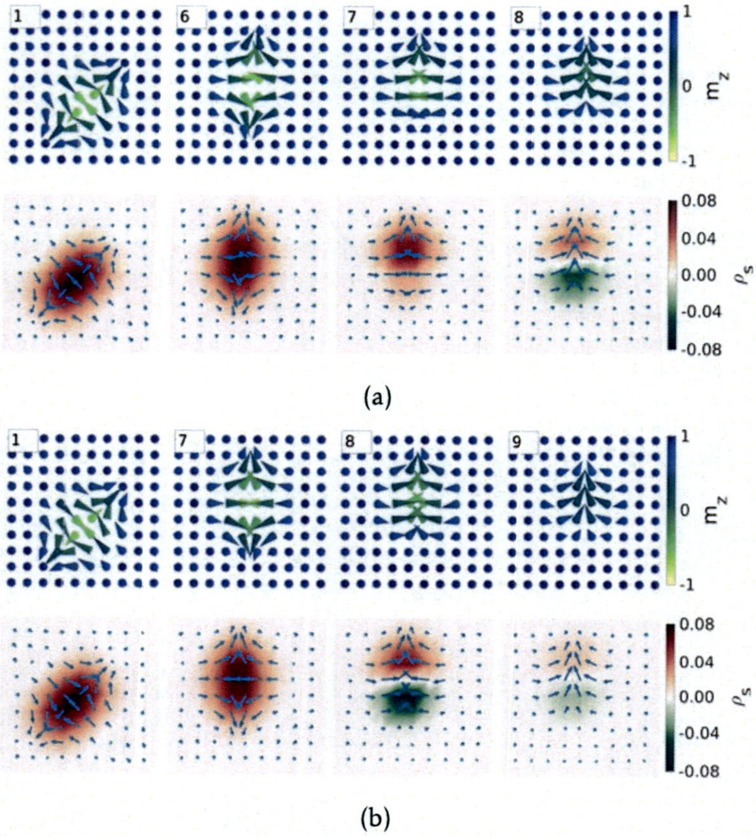

Fig. 6.4 Spin maps (zoomed) and corresponding topological charge density along the transition path for an antiskyrmion annihilation with $(b, d) = (0.2, 0.03)$ where **a** shows the path over SP_1 and **b** shows the path over SP_2. The image index is given in the top left-hand corner. The antiskyrmion annihilates via the nucleation of a half skyrmion, or meron. The saddle point is im. 7 in both cases and precedes the meron nucleation, which occurs at im. 8

order saddle point, for which we show spin maps at selected images in Fig. 6.4a and b, where the corresponding energy profiles are shown in Fig. 6.5a and b. In both cases, the annihilation process begins with the rotation of the helicity of the antiskyrmion, such that its Néel-type axes are aligned along the first-neighbour axes (ims. [1-7]). This configuration constitutes an energy maximum as it is unfavoured by the DMI, and the antiskyrmion can reach this state either as a full antiskyrmion with full integer topological charge (SP_2 at im. 7 in Figs. 6.4b and 6.5b), or the spins may begin to unwind, which leads to a half-integer topological charge (SP_1 at im. 7 in Figs. 6.4a and 6.5a). Past the saddle point, similarly to the annihilation mechanism of its skyrmion counterpart, the topological charge drops to zero as a meron is nucleated (im. 8) and annihilates with the remaining antimeron (ims. [9-10]). SP_1 exists in four possible realizations ($\pi/2$-rotations), while SP_2 exists in two

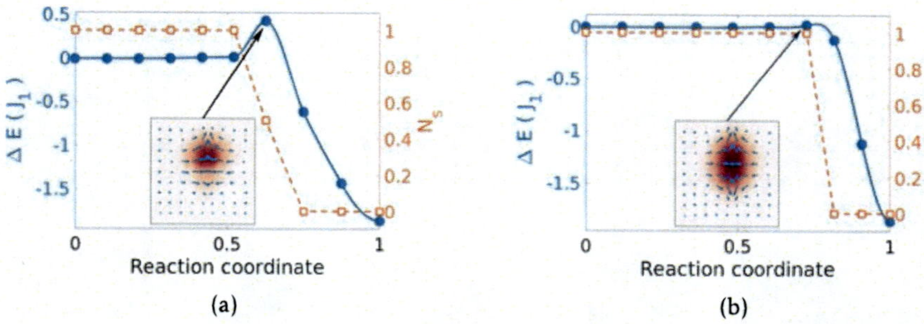

Fig. 6.5 Internal energy barrier in units of J_1 (filled blue dots) and topological charge N_s (unfilled red squares) along the transition path for an antiskyrmion annihilation with $(b, d) = (0.2, 0.03)$ for **a** the path over SP_1, **b** the path over SP_2. The inset shows the spin configuration and the topological charge density at the saddle point. SP_1 carries a 1/2 topological charge, while SP_2 carries a full integer charge

(π-rotation). Therefore, even though the barrier associated with SP_2 appears very small, it is necessary to compute an attempt frequency before drawing any conclusion on whether or not the mechanism involving SP_1 should contribute to the total rate. It is interesting to note that the fact that the antiskyrmion displays frustration in its helicity suffices to distinguish between a metastable solution and a saddle point. In other words, the saddle point SP_2 corresponds to the same configuration as the metastable antiskyrmion state, but with a $\pi/4$-rotation of helicity.

6.3.3 Second-Order Skyrmion

We select parameters that allow both the first- and second-order skyrmions to be metastable: $(b, d) = (0.14, 0.005)$. We initially set the final state of the GNEB calculation to a Néel skyrmion, and we obtain the mechanism in Fig. 6.6a, with the corresponding energy and topological charge profiles shown in Fig. 6.7a. The second-order skyrmion first undergoes a rotation of helicity (ims. [1-3]), and reaches the saddle point (im. 3). The nucleation of an antimeron along the Néel axis follows at image 4, at which point the energy drops, and the topological charge goes from $N_s = -2$ to $N_s = -1 + \frac{1}{2} - \frac{1}{2} = -1$. The antimeron subsequently annihilates with part of the second-order skyrmion, leaving a first-order Néel skyrmion in its place (ims. [4–8]).

Another route to annihilation for the second-order skyrmion corresponds to its division into a bound skyrmion pair, as shown in Fig. 6.6b. The corresponding energy and topological charge profiles are shown in Fig. 6.7b. The bound skyrmion pair is a metastable solution, since the interaction potential of skyrmions in frustrated magnets is non-monotonic as a function of distance, and is found in turn positive (repulsive) and negative (binding) [5, 18]. This behaviour is not present in chiral magnets, in which the interaction between skyrmions is always repulsive. This mechanism does not involve any significant change in the topological charge of the total system,

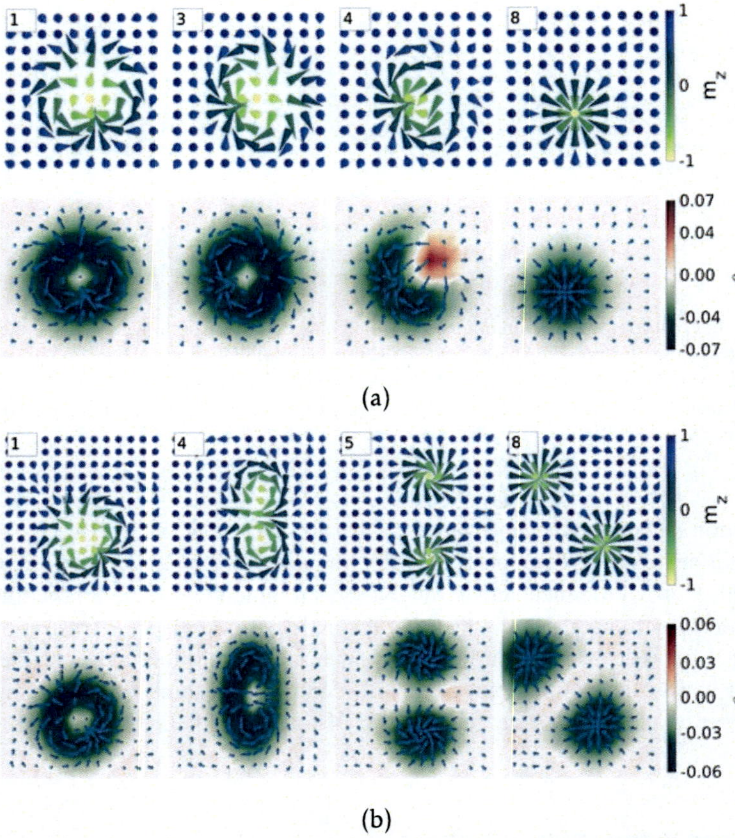

Fig. 6.6 Spin maps (zoomed) and corresponding topological charge density along the transition path for **a** the decay of a second-order skyrmion into a first-order skyrmion via antimeron nucleation, with $(b, d) = (0.14, 0.005)$, **b** the division of a second-order skyrmion into a bound skyrmion pair, with $(b, d) = (0.1, 0.005)$. The image index is given in the top left-hand corner. In **a**, the SP is im. 3 and preceeds the antimeron nucleation. In **b**, the SP is im. 4 and preceeds the separation of the two skyrmions

but instead results in a redistribution of the topological charge density. The saddle point in this case is the state preceeding the separation of the two skyrmions (im. 4). Both of the above mechanisms possess four realizations, corresponding to $\pi/4$ rigid rotations of the system. As for the fate of the skyrmion pair, the binding potential is quite shallow compared to the activation barrier for annihilation (from the results of Ref. [18], we can estimate the unbinding barrier to be $\sim 10^{-4}\ J_1$). Therefore, the skyrmions will, in most cases, separate before annihilating individually.

Similar processes were reported in Ref. [21] from dynamics simulations, in which the second-order skyrmion division was triggered by both a driving current at zero temperature, and thermal fluctuations at finite temperature. The finite temperature

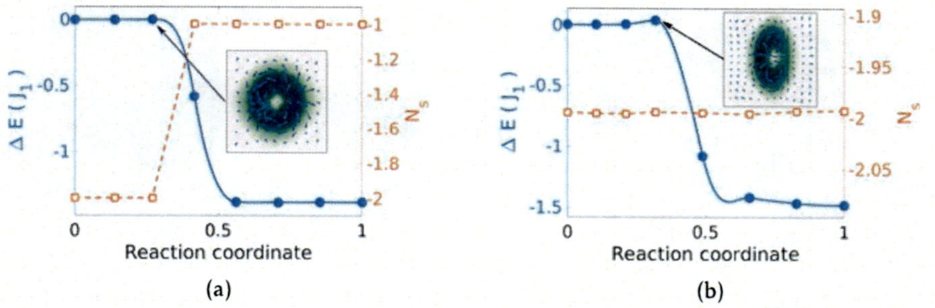

Fig. 6.7 Internal energy barrier in units of J_1 (filled blue dots) and topological charge N_s (unfilled red squares) along the transition path for **a** the decay of a second-order skyrmion into a first-order skyrmion via antimeron nucleation, with $(b, d) = (0.14, 0.005)$, **b** the division of a second-order skyrmion into a bound skyrmion pair, with $(b, d) = (0.1, 0.005)$. The inset shows the spin configuration and the topological charge density at the saddle point

simulations also showed the decay of the second-order skyrmion into a first-order skyrmion.

6.3.4 (Anti)meron Nucleation and Evolution of the Topological Charge

In order to visualize the evolution of the topological charge, we show the "ball in a net" representations of some annihilation mechanisms presented above in Fig. 6.8. As previously described in Sect. 2.3.2, vertices correspond to the tip of the magnetic vectors with their origin in the centre of the sphere, and the edges represent the exchange coupling between first neighbours. The ferromagnetic background (spins along $+z$) is found at the north pole of the sphere, while the core of a skyrmion typically points along the south pole (along $-z$). The view is set just below the south pole, looking axially towards the $+z$ direction, so that the core of the skyrmion is found at the centre of the view.

We show a skyrmion annihilation via antimeron nucleation (Fig. 6.8a) and isotropic collapse (Fig. 6.2b), an antiskyrmion annihilation via meron nucleation (Fig. 6.4b, which is the path that traverses SP_1), and the decay of a second-order skyrmion via antimeron nucleation (Fig. 6.8d). Annihilations that occur via the injection of the opposite topological charge (Fig. 6.8a and c) involve a rearrangement of the spin-mesh in a way that allows the entire net to be removed along a single direction. By contrast, an isotropic collapse consists in progressively spreading the mesh open by an equal amount along all directions (Fig. 6.8b). Lastly, a second-order skyrmion corresponds to a spin configuration in which the magnetization wraps the unit sphere twice, therefore, in Fig. 6.8d, it looks like the ball is inside a double net. The decay into a first-order skyrmion via the injection of an antimeron consists in the removal of one of the nets, again along a given direction, so that only a single net remains.

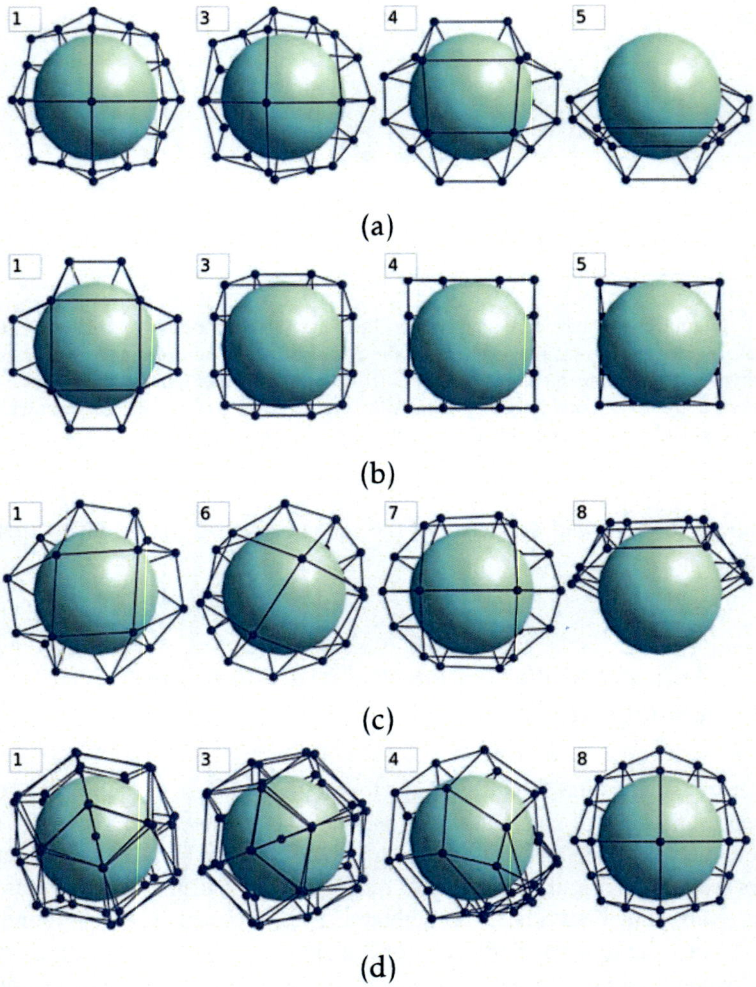

Fig. 6.8 "Ball in a net" representation of the spin configuration at selected images along the path to annihilation. The image index is given in the top left-hand corner. We show **a** a skyrmion annihilation via antimeron nucleation (Fig. 6.2a), **b** a skyrmion annihilation via isotropic collapse (Fig. 6.2c), **c** an antiskyrmion annihilation via meron nucleation (Fig. 6.4a), **d** a second-order skyrmion decay into a first-order skyrmion (Fig. 6.6a)

6.4 Langevin Simulations

In order to validate the previous paths calculated with GNEB, we perform direct Langevin simulations at low temperature. Details on the method are given in Sect. 5.2. The values of the parameters are the same as in Chaps. 4 and 5: $J_1 = 1.6 \times 10^{-20}$ J, $\mathscr{M}_s = 1.1 \times 10^6$ A m^{-1}, $\alpha = 0.5$, and $\gamma_e = 1.76 \times 10^{11}$ rad s^{-1} T^{-1}. We record

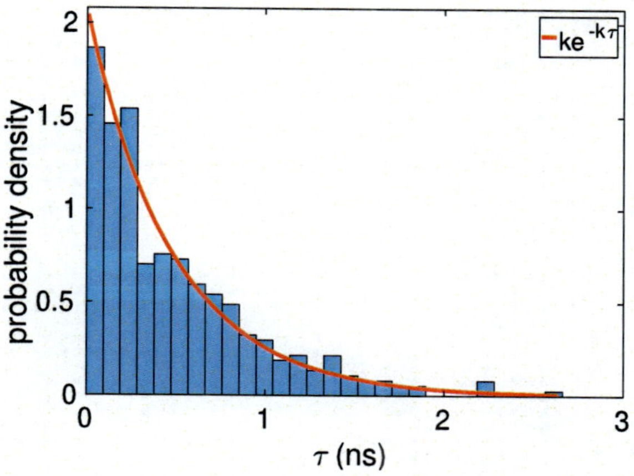

Fig. 6.9 Probability density distribution of the lifetimes of the antiskyrmion computed over 390 direct Langevin simulations at $T = 80\,\text{K}$. The red curve is the theoretical PDF obtained with $k = k_{\text{Langer}}$ given in Eq. (6.1)

annihilation mechanisms for which we show snapshots of the spin configuration in Fig. 6.10.

For the antiskyrmion, at $T = 80\,\text{K}$ and $(b, d) = (0.2, 0.03)$, which we show in Fig. 6.10a, we observe the meron-nucleation type of annihilation that we reported in Fig. 6.4. Since we have short lifetimes within the nanosecond time scale, we can estimate the antiskyrmion's stability directly from Langevin simulations. We use the topological charge of the system to track the collapse of the antiskyrmion, that is, a collapse is recorded when $N_s = 0$. We record 390 collapses and obtain an average lifetime $\tau_{\text{Lgv}}(80\text{ K}) = 0.49$ ns, with a standard deviation of the order of the mean lifetime, $\sigma = 0.46$ ns. The fact that the standard deviation is approximately the same as the mean value is coherent with the properties of an exponentially distributed variable such as τ. Figure 6.9 shows the probability density distribution of the computed lifetimes, which matches the probability density function (PDF) $f(k, \tau) = ke^{-k\tau}$ plotted in red, where we have used $k = k_{\text{Langer}}$ given in Eq. (6.1).

As for the second-order skyrmion and antiskyrmion, as they overall exhibit lower activation energies, we perform simulations at $T = 50\,\text{K}$, and $(b, d) = (0.14, 0.005)$. The division of the second-order skyrmion into a bound skyrmion pair seems to occur along at least two different paths that we show in Fig. 6.10b and c, in which the latter corresponds to the MEP in Fig. 6.6b. In Fig. 6.10b, the division occurs through excitation of one half of the second-order skyrmion, which eventually leads to the separation of the two halves. On the other hand, in Fig. 6.10c, we witness a more symmetric kind of division. For the same set of parameters, we also observe the decay into a single skyrmion shown in Fig. 6.10d, which corresponds to the path in Fig. 6.6a. Finally, in Fig. 6.10e, we show the division of a second-order antiskyrmion into a pair of first-order antiskyrmions. Since the pairwise interaction potential between antiskyrmions depends on their relative helicities in addition to their distance [18], this mechanism is difficult to relax with GNEB due to the overall complexity of the energy landscape, so we only observed this particular mechanism via Langevin simulations.

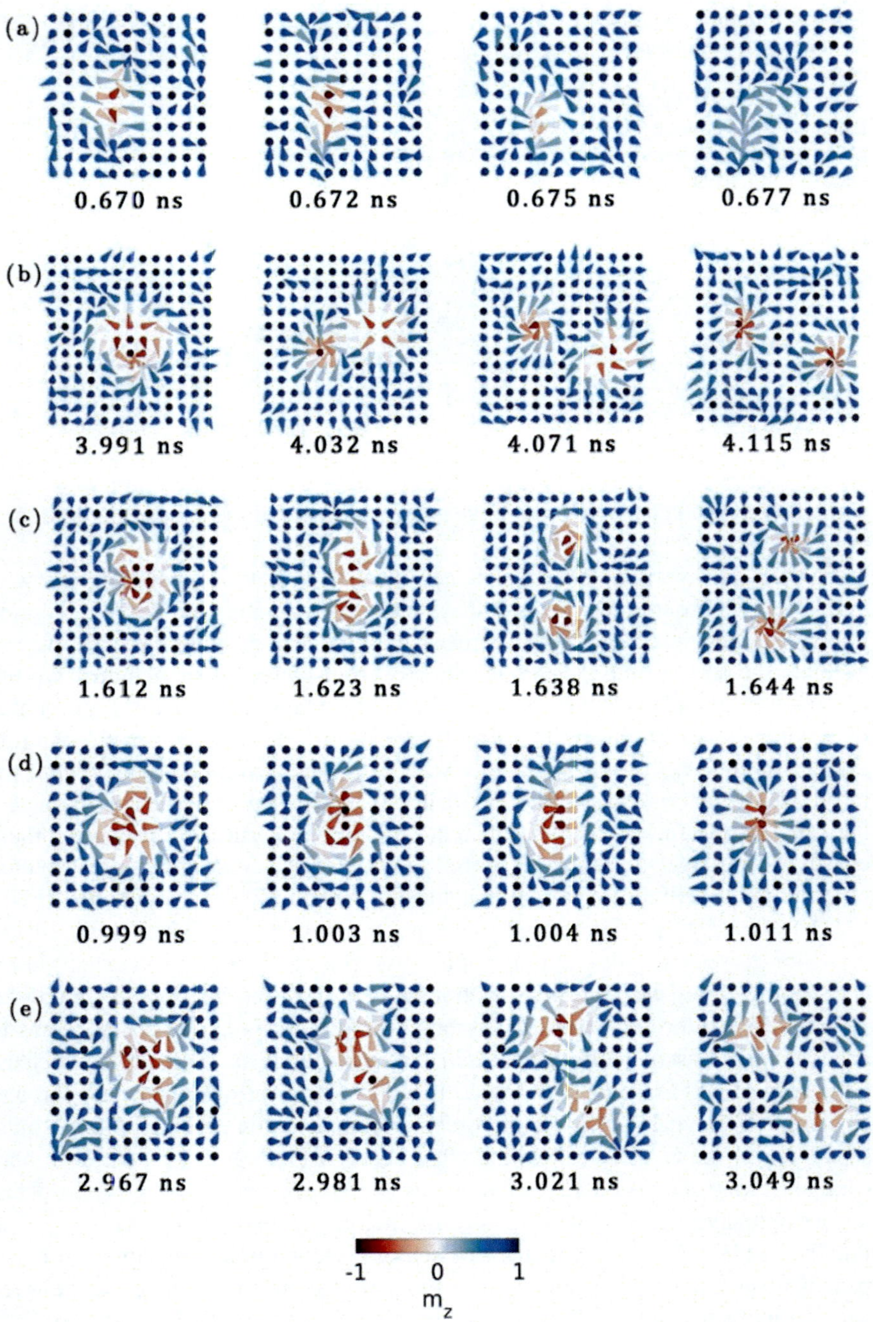

Fig. 6.10 Langevin simulation snapshots (zoomed) of **a** an antiskyrmion annihilation, **b**, **c** the division of a second-order skyrmion into a bound skyrmion pair, **d** the decay of a second-order skyrmion into a first-order skyrmion, **e** the division of a second-order antiskyrmion into a bound antiskyrmion pair. The antiskyrmion annihilation is shown at $T = 80\,\mathrm{K}$, $(b, d) = (0.2, 0.03)$, while the other simulations are carried out at $T = 50\,\mathrm{K}$, $(b, d) = (0.14, 0.005)$

6.5 Thermal Stability

We use Langer's theory to compute a complete rate for each of the mechanism presented above. The results are gathered in Table 6.1. To account for all equivalent realizations of a given mechanism (rigid rotations of the spin configurations at the saddle point), the prefactor f_0 is multiplied by a factor two or four when it is appropriate, in accordance with Sect. 6.3. In the case of the decay of the second-order skyrmion into a first-order skyrmion, we find a Goldstone mode at both the metastable and transition states, which do not seem to clearly correspond to a translational or rotational invariance, hence we cannot give a definite value for the attempt frequency in this case. Based on these results, we can estimate the lifetime of the antiskyrmion calculated with Langer's theory, and compare it with the average lifetime previously obtained from direct Langevin simulations. Assuming independent processes, we have

$$\begin{aligned} k_{\text{Langer}}(80\text{ K}) &= f_{01}e^{-\beta\Delta E_1} + f_{02}e^{-\beta\Delta E_2}, \\ &= \tau_{\text{Langer}}^{-1}(80\text{ K}), \\ &= 2.08\text{ GHz}, \end{aligned} \tag{6.1}$$

and $\tau_{\text{Langer}}(80\text{ K}) = 0.48$ ns. Although we are not completely within the scope of Langer's theory ($\beta\Delta E_1 = 0.3$, as opposed to the recommended $\beta\Delta E \geq 5$ [22]), this shows an excellent agreement with the result from direct Langevin, $\tau_{\text{Lgv}}(80\text{ K})$ = 0.49 ns (see also the agreement of the probability density in Fig. 6.9).

In Fig. 6.11, we show the average lifetimes of the skyrmion and the antiskyrmion as a function of temperature with respect to the different mechanisms. Overall, the structures are much less stable than the chiral solutions in Chap. 4. The average lifetimes at room temperature are typically less than a nanosecond. This is not surprising, because the barriers are small compared to the exchange strength ($\Delta E_{\text{sk}} < 2J_1$ for

Table 6.1 Internal energy barrier, ΔE, and rate prefactor, f_0, for the annihilation of a skyrmion, an antiskyrmion, and a second-order skyrmion with different values of the reduced applied field b and reduced DMI d

Mechanism	b (J_1)	d (J_1)	ΔE (J_1)	f_0 (GHz)
		Skyrmion		
Antimeron nucleation	0.2	0.03	1.27	109.9 × 4
Isotropic collapse	0.3	0.07	1.62	3639.5
Isotropic collapse	0.7	0.2	1.80	1247.5
		Antiskyrmion		
Meron nucleation SP_1	0.2	0.03	0.43	10.6 × 4
Meron nucleation SP_2	0.2	0.03	0.022	1.4 × 2
2nd. order sk.				
Antimeron nucleation	0.14	0.005	0.0011	–
Division	0.14	0.005	0.062	–

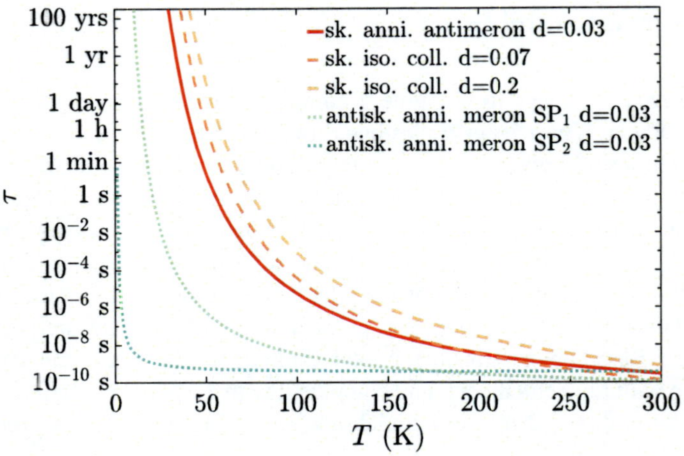

Fig. 6.11 Average lifetime of a skyrmion and an antiskyrmion under frustrated exchange as a function of temperature with respect to all considered mechanisms calculated via Langer's theory as given in Eq. (3.31). $k = 0.1$ in all cases. The value of b, the energy barriers and the value of the prefactor are given in Table 6.1

the skyrmion and $\Delta E_{\text{antisk}} < 1J_1$ for the antiskyrmion), and the attempt frequencies are of the order of 1–1000 GHz. The more stable skyrmion is the skyrmion at higher DMI ($d = 0.2$) which annihilates via isotropic collapse. This is because it has the highest barrier, and second lowest attempt frequency. At 300 K, its lifetime is 0.84 ns. At low temperature, the least stable skyrmion is the one annihilating via antimeron nucleation at low DMI ($d = 0.03$), although, at room temperature, even though it has the lowest energy barrier, it is more stable than the solution at intermediate DMI ($d = 0.07$) which undergoes a collapse with a small asymmetry. This is because the attempt frequency associated with meron nucleation is one order of magnitude smaller than the ones associated with collapse mechanisms. The antiskyrmion also provides an interesting example where the attempt frequency plays an essential role. The energy barrier associated with SP_2 is very small compared to the exchange strength (0.022 J_1) and much smaller than the one associated with SP_1, so one would typically assume that the mechanism involving SP_2 should dominate the rate. However, the attempt frequency associated with SP_2 is not only ten times lower than the one associated with SP_1, but SP_2 can be realized in two different ways, i.e., whenever the helicity of the antiskyrmion is 0 or π, while SP_1 can be realized in four different ways, that is, the inset in Fig. 6.4a and three $\pi/4$ rotations. It follows that the total attempt frequency associated with SP_1 is 15 times larger than the one associated with SP_2. Consequently, SP_2 dominates the rate at very low temperature ($T \ll 100$ K), but the lifetimes intersect around 175 K. Beyond that limit, the mechanism involving SP_1 is the most probable one. Even at 80K, the contribution from both saddle points is necessary for Langer's result to match the average lifetime and the probability density distribution of the lifetimes obtained from direct Langevin simulations (see Fig. 6.9).

6.6 Discussion

We have seen that, under frustrated exchange couplings, and in regions of parameter space for which both metastable skyrmion and antiskyrmion solutions can be realized, the most probable paths to annihilation appear to be the paths involving the injection of the opposite topological charge into the system, by means of the nucleation of merons and antimerons. The injection of the opposite charge usually happens shortly after the system has passed the saddle point configuration, or at the saddle point in the case of the skyrmion, and also coincides with the loss of one unit of topological charge. The state preceding the injection of the opposite charge may also possess a half-integer topological charge, as is the case of the antiskyrmion in the annihilation path involving the saddle point SP_1 (im. 7 in Figs. 6.4a, 6.5a, and 6.8c).

Overall, the skyrmionic solutions in this system were found to be rather unstable, as they possess low internal energy barriers compared to the exchange strength, combined with many possible paths to annihilation with attempt frequencies in the range of several gigahertz to several hundreds of gigahertz. The existence of many possible paths is a direct consequence of the exchange frustration, which drastically complexifies the energy landscape and is responsible for the emergence of many (meta)stable and saddle point states, as well as many possible paths connecting them. At room temperature, the average lifetimes were calculated to be under a nanosecond. We conclude that this specific set of exchange parameters on the square lattice derived in Ref. [18] does not seem to yield isolated skyrmion solutions that could potentially be used in applications. In the present case, it seems that the skyrmion stabilized at higher DMI which undergoes isotropic collapse is a more stable solution than the low DMI solution whose annihilation occurs via antimeron nucleation. Additionally, the low DMI skyrmion might also undergo isotropic collapse. This mechanism is associated with a higher-order saddle point which we did not relax here, and it would contribute to the reduction of the skyrmion's lifetime. From there, one may conclude that the low DMI region of parameter space where metastable skyrmions are primarily stabilized by exchange frustration and antiskyrmions are also metastable solutions is not a region where isolated skyrmions will be the most stable, and so the existence of antiskyrmions should be an indicator of lower stability of the skyrmions. However, the attempt frequency of the meron nucleation is one order of magnitude lower than that of isotropic collapse processes. Therefore, as is often the case, things are not that simple, and the above conclusions only apply to the specific set of parameters we have studied in this chapter. In Ref. [19], we concluded that skyrmions undergo annihilation via antimeron injection in the region of parameter space where low DMI allows antiskyrmions to coexist with skyrmions as metastable solutions. With the selected set of parameters, we also had to increase the applied field with increasing DMI in order to be able to relax isolated (anti)skyrmion solutions (see the position of the orange dots in Fig. 6.1). At higher DMI and applied field, antiskyrmion solutions disappear and we recovered an isotropic type of collapse. In Ref. [4], small skyrmions were observed at zero field in Rh/Co/Ir(111) via SP-STM. A strong exchange frustration in this system was found from DFT calculations. GNEB simulations revealed that in Rh/Co/Ir(111), skyrmions at zero field undergo a similar annihilation to the one we have reported in this chapter, involving the injection of the opposite topological charge—which the authors referred to as "chimera" [8] collapse. At higher applied field, the isotropic collapse is favoured. This conclusion agrees partially with the results of this chapter. We did not find here a limit where, by increasing the field at constant DMI, the annihilation mechanism would change, however this still may be the case in a very small window of increasing applied field before skyrmions are no longer solutions, in which case it is possible that we missed it. Instead, we had to increase both the field and DMI to recover the isotropic collapse. The existence of antiskyrmions was not mentioned in Ref. [4]. We note that the system studied by the authors of Ref. [4] used a Co magnetic layer and sub-10-nm skyrmions were reported in the FM state at zero field. This configuration resembles the type of stabilization conditions of we have studied in Chap. 4 for DMI stabilization, and in the present chapter for the effect of exchange frustration. In all cases we used material parameters of Pt/Co/AlO_x for J_1 and the saturation magnetization. The energy barriers we have found here for the skyrmion are of a few hundreds of meV (127–180 meV), while in [4], they were calculated to be around 250–290 meV, which is similar to the barriers we reported for chiral skyrmion collapse in Chap. 4. This hints at a potential higher stability of the skyrmions in Rh/Co/Ir(111) than what we have reported in this chapter. However, a rate prefactor was not computed in Ref. [4].

The low stability of the studied solutions allowed for a direct comparison of the average lifetime of the antiskyrmion at 80 K computed from direct Langevin simulations with Langer's theory calculations, and we obtained a very good agreement. Nevertheless, in this context, the direct use of reaction rate theory to estimate an overall lifetime for any given structure seems ill-advised and unreliable, as it is difficult to account for all possible mechanisms. As for the low activation energies and not particularly low attempt frequencies, we can once more relate them to the skyrmions being very small, and possessing only a few internal modes [11, 23]. Lastly, this chapter yet again contained some interesting examples where the attempt frequency plays an important role in the stability, especially in the higher temperature region. The most striking case is that of the antiskyrmion, which annihilates in this system via meron nucleation. One of its annihilation mechanisms involves a very small activation barrier of 0.022 J_1 at saddle point SP_2, yet, above a temperature of 175 K, this mechanism yields a longer lifetime that the other mechanism over SP_1, which corresponds to an activation energy of 0.43 J_1. Even at 80K, we found that both mechanisms should be taken into account in Langer's theory calculations in order to match the results of direct Langevin simulations.

References

1. Romming N, Hanneken C, Menzel M, Bickel JE, Wolter B, von Bergmann K, Kubetzka A, Wiesendanger R (2013) Writing and deleting single magnetic skyrmions. Science 341(6146):636–639

2. Dupé B, Hoffmann M, Paillard C, Heinze S (2014) Tailoring magnetic skyrmions in ultra-thin transition metal films. Nat Commun 5:4030
3. Romming N, Pralow H, Kubetzka A, Hoffmann M, von Malottki S, Meyer S, Dupé B, Wiesendanger R, von Bergmann K, Heinze S (2018) Competition of Dzyaloshinskii-Moriya and higher-order exchange interactions in Rh/Fe atomic bilayers on Ir(111). Phys Rev Lett 120:207201
4. Meyer S, Perini M, von Malottki S, Kubetzka A, Wiesendanger R, von Bergmann K, Heinze S (2019) Isolated zero field sub-10 nm skyrmions in ultrathin Co films. Nat Commun 10:3823
5. Leonov A, Mostovoy M (2015) Multiply periodic states and isolated skyrmions in an anisotropic frustrated magnet. Nat Commun 6:8275
6. Okubo T, Chung S, Kawamura H (2012) Multiple-q states and the skyrmion lattice of the triangular-lattice Heisenberg antiferromagnet under magnetic fields. Phys Rev Lett 108(1):017206
7. von Malottki S, Dupé B, Bessarab P, Delin A, Heinze S (2017) Enhanced skyrmion stability due to exchange frustration. Sci Rep 7(1):12299
8. Rózsa L, Palotás K, Deák A, Simon E, Yanes R, Udvardi L, Szunyogh L, Nowak U (2017) Formation and stability of metastable skyrmionic spin structures with various topologies in an ultrathin film. Phys Rev B 95(9):094423
9. Ritzmann U, Von Malottki S, Kim J-V, Heinze S, Sinova J, Dupé B (2018) Trochoidal motion and pair generation in skyrmion and antiskyrmion dynamics under spin-orbit torques. Nat Electron 1(8):451
10. Bessarab PF, Müller GP, Lobanov IS, Rybakov FN, Kiselev NS, Jónsson H, Uzdin VM, Blügel S, Bergqvist L, Delin A (2018) Lifetime of racetrack skyrmions. Sci Rep 8(1):1–10
11. Desplat L, Suess D, Kim J-V, Stamps RL (2018) Thermal stability of metastable magnetic skyrmions: entropic narrowing and significance of internal eigenmodes. Phys Rev B 98:134407
12. Bessarab PF, Uzdin VM, Jonsson H (2015) Method for finding mechanism and activation energy of magnetic transitions, applied to skyrmions and antivortex annihilation. Comput Phys Commun 196:335–347
13. Lobanov IS, Jónsson H, Uzdin VM (2016) Mechanism and activation energy of magnetic skyrmion annihilation obtained from minimum energy path calculations. Phys Rev B 94(17):174418
14. Uzdin VM, Potkina MN, Lobanov IS, Bessarab PF, Jónsson H (2017) The effect of confinement and defects on the thermal stability of skyrmions. Phys B: Condens Matter 549:6–9
15. Cortés-Ortuño D, Wang W, Beg M, Pepper RA, Bisotti M-A, Carey R, Vousden M, Kluyver T, Hovorka O, Fangohr H (2017) Thermal stability and topological protection of skyrmions in nanotracks. Sci Rep 7(1):4060
16. Stosic D, Mulkers J, Van Waeyenberge B, Ludermir TB, Milošević MV (2017) Paths to collapse for isolated skyrmions in few-monolayer ferromagnetic films. Phys Rev B 95(21):214418
17. Heil B, Rosch A, Masell J (2019) Universality of annihilation barriers of large magnetic skyrmions in chiral and frustrated magnets. Phys Rev B 100:134424
18. Lin S-Z, Hayami S (2016) Ginzburg-Landau theory for skyrmions in inversion-symmetric magnets with competing interactions. Phys Rev B 93(6):064430
19. Desplat L, Kim J-V, Stamps RL (2019) Paths to annihilation of first-and second-order (anti) skyrmions via (anti) meron nucleation on the frustrated square lattice. Phys Rev B 99(17):174409
20. Henkelman G, Uberuaga BP, Jónsson H (2000) A climbing image nudged elastic band method for finding saddle points and minimum energy paths. J Chem Phys 113(22):9901–9904
21. Zhang X, Xia J, Zhou Y, Liu X, Zhang H, Ezawa M (2017) Skyrmion dynamics in a frustrated ferromagnetic film and current-induced helicity locking-unlocking transition. Nat Commun 8(1):1717
22. Coffey W, Garanin D, McCarthy D (2001) Crossover formulas in the Kramers theory of thermally activated escape rates- application to spin systems. Adv Chem Phys 117:483–765
23. von Malottki S, Bessarab PF, Haldar S, Delin A, Heinze S (2019) Skyrmion lifetime in ultrathin films. Phys Rev B 99(6):060409

Chapter 7
Conclusion and Outlook

In this thesis, we have studied the stability of metastable magnetic skyrmions on the two-dimensional square lattice. This task was firstly carried out via a numerical implementation of Kramers' method for a multidimensional energy landscape set in the intermediate-to-high damping regime, i.e., Langer's statistical theory for the decay of metastable states. The paths of minimum energy that lead to the skyrmion annihilation were computed via the geodesic nudged elastic band method. Chiral magnetic skyrmions stabilized at zero magnetic field by the Dzyaloshinskii-Moriya interaction were found to annihilate via two distinct isotropic collapse mechanisms, respectively passing through a first-, and a third-order saddle point on the energy landscape. In both cases, the loss of the topological charge occurs past the saddle point. The contribution of the third-order saddle point to the rate was found to be negligible, which justified considering only paths through first-order saddle points in the rest of this work. Alternatively, the skyrmions may diffuse to the edge of the system and escape through the boundary. The computation of a rate prefactor revealed that the thermally significant eigenmodes are the modes localized to the skyrmion, in contrast to the rest of the collective spin-wave modes, which extend to the entire spin lattice and contribute weakly. Important variations of the rate prefactor, also known as the attempt frequency, were found between different mechanisms on the one hand, and between different values of the destabilizing applied magnetic field on the other. These variations originate from strong entropic effects, which come from the difference in configurational entropy between the metastable skyrmion state and the saddle point. In the model system we studied, the configurational entropy was found to decrease between the metastable state and the saddle point ($\Delta S < 0$), which results in lowered attempt frequencies. This therefore constitutes a stabilizing effect, which we have referred to as *entropic narrowing in the saddle point region*. The strong entropic contribution mainly originates from the skyrmion's internal modes, and is generally more pronounced for collapse mechanisms. One important consequence of this result is that, at room temperature, the most probable mechanism is

L. Desplat, *Thermal Stability of Metastable Magnetic Skyrmions*, Springer Theses,
https://doi.org/10.1007/978-3-030-66026-0_7

not necessarily the one with the lowest internal energy barrier, because it may also possess a low attempt frequency. We used material parameters of Pt/Co/AlO$_x$ thin films, in which we tuned the effective anisotropy and strength of the DMI to stabilize isolated skyrmions of about 10 nm in diameter at zero applied magnetic field. The activation barriers were calculated to be around 300 meV. At room temperature, the skyrmion was found to be more stable with respect to collapse mechanisms, due to uncharacteristically low attempt frequencies for collapse in the megahertz regime. Far away from edges and structural defects, the average lifetime of the skyrmion against collapse was found to be of the order of a few milliseconds.

Next, we used forward flux sampling, which is a path sampling method for the simulation of rare events, to compute collapse rates of metastable skyrmions in an infinite system. The goal was to compare the FFS results with the ones obtained from Langer's theory. This was an important step, because the use of Langer's theory is based on many assumptions. An overall good agreement was found between Langer's theory and FFS, thus confirming the strong dependence of the attempt frequency on the external magnetic field. Within our range of studied applied field values ($0 \leq B \leq$ 730 mT), the attempt frequency for the collapse extends from a few megahertz at zero-field, to a few gigahertz—or a few tens of gigahertz—at higher field, thus spanning over three to four orders of magnitude. This leads to the following conclusion: *due to a strong entropic contribution, internal energy barriers are not enough in order to correctly predict the lifetime of magnetic skyrmions, and it is essential to also evaluate a rate prefactor.* This constitutes the main result of the present thesis.

Lastly, we looked at paths to annihilation of first- and second-order skyrmions and antiskyrmions on the frustrated $J_1 - J_2 - J_3$ square lattice. Frustrated exchange has been shown to arise at certain transition metal interfaces by DFT calculations, most notably in Pd$_{fcc}$/Fe, Rh/Fe, and Rh/Co bilayers on Ir(111). We found that, under frustrated exchange couplings, and in certain regions of parameter space, the annihilation of skyrmionic solutions no longer occurs through an isotropic type of collapse, and instead involves the injection of the opposite topological charge into the system, by means of the nucleation of merons and antimerons, which carry half-integer tolopological charges. We added a weak interfacial DMI into the system to break degeneracies, and we found that the antimeron injection was a preferred mechanism for the annihilation of skyrmions in the low DMI and low magnetic field region of parameter space, where antiskyrmion solutions also exist. The injection of the opposite charge usually happens shortly after the system has passed the saddle point configuration, or at the saddle point in the case of the skyrmion, and also coincides with the loss of one unit of topological charge. Alternatively, the second-order (anti)skyrmion may split into a bound (anti)skyrmion pair, which involves no change in the total topological charge. In all cases, the skyrmions in this particular system only span over a few lattice sites and were found to be quite unstable, both in terms of activation energy and rate prefactor. By using the first-neighbour exchange coupling constant and saturation magnetization of Pt/Co/AlO$_x$, internal energy barriers were calculated to be below 200 meV for the skyrmion—or below 50 meV for the antiskyrmion—and the prefactor was found in the range of a few gigahertz to a few thousands of gigahertz. In all cases, the (anti)skyrmion lifetimes at room temperature were under one nanosecond. The most stable skyrmion was found to be the one stabilized at high DMI which annihilates via isotropic collapse. For the system we studied, this implies that the existence of antiskyrmions may be an indicator of lower skyrmion stability. This is not necessarily true in other systems. For instance, in Ref. [1], higher internal energy barriers were typically associated with the antimeron injection, which is preferred over the isotropic collapse at low applied field. However, a prefactor was not computed in Ref. [1].

The existence of a strong entropic contribution to the thermal stability of skyrmions was initially proposed in [2], but the treatment only compared the entropy of the skyrmion state to that of the ferromagnetic state without considering the saddle point configuration, and was therefore incomplete. The following works mentioned below were published over the duration of this thesis. Experimentally, strong entropic effects were observed by Wild et al. in the case of the decay of the skyrmion lattice in the B20 chiral magnet $Fe_{1-x}Co_xSi$ [3]. The authors measured a variation of the Arrhenius prefactor over 30 orders of magnitude when the applied field changed by about 20 mT, and attributed this effect to a case of entropy-enthalpy compensation, also known as the Meyer-Neldel rule [4, 5]. Meanwhile, a similar methodology to the one presented in this thesis was used by Bessarab et al. [6], in which the rate prefactor was calculated by applying harmonic transition state theory [7]. The aforementioned work proposed a treatment of the translational Goldstone modes of the skyrmion in a case where the metastable skyrmion state decouples from the lattice, which yields a temperature-dependence of the prefactor, as well as a system size-dependence for escape processes. This was not treated in this thesis, in which the skyrmions are not large compared to the lattice constant, and so the coupling of the skyrmion to the underlying lattice breaks the translational symmetry, and leads to lattice effects in the rate prefactor. In Ref. [8], we showed that the treatment of the translation modes of the skyrmions studied in the present thesis as Goldstone modes leads to an overestimation of the collapse prefactor by up to one order of magnitude at zero magnetic field.

In line with the above discussion, the different choice of material parameters in Ref. [6], which was carried out via ab initio density functional theory (DFT) calculations for Pd/Fe/Ir(111), yielded attempt frequencies and skyrmion lifetimes vastly different from the ones reported here, with attempt frequencies for collapse spanning from tens of gigahertz, to hundreds of terahertz. From what we understand, HTST is similar to Langer's theory and also pertains to the intermediate-to-high damping (IHD) regime, and the main difference between the theories is the calculation of the dynamical contribution to the prefactor (λ_+ in Langer's theory). The prefactor in HTST does not explicitly depend on the damping value, and should therefore give an upper bound of the attempt frequency over the range of large possible dampings. Interestingly, we have found that skyrmion collapse rates calculated with Langer's theory are not strongly affected by the choice of damping, although the theory is, of course, only applicable to the IDH regime. Apart from collapse and escape processes, duplication and merging of skyrmions were also reported as possible transition mechanisms by a mode following method [9]. Lastly, Ref. [10] reported a drastic change in the attempt frequency of up to seven orders of magnitude per change in magnetic field of one tesla for skyrmion collapse in hcp − Pd/Fe/Ir(111), which is consistent with the main results of this thesis. The authors found that, above a certain size of the metastable skyrmion, the entropy difference between the skyrmion and the saddle point seems to decrease linearly (become more negative) with the surface area r^2 of the skyrmion. For very small skyrmions ($r^2 \lessapprox 6\text{nm}^2$), $\Delta S > 0$, thus leading to a destabilizing effect. This confirms the somewhat intuitive result that smaller skyrmions are less stable, which we have also seen in Chap. 4 by varying the destabilizing magnetic field. This can be understood by the fact that the internal energy barrier decreases with the skyrmion size (fewer spins to reverse), and, concurrently, the attempt frequency increases under the effect of the entropic contribution.

The potential use of skyrmions in magnetic data storage [11, 12] requires sub-10-nm skyrmions to be highly stable at room temperature in order to compete against existing technology such as domain wall racetrack memories. Despite the somewhat pessimistic conclusions drawn above, some other factors may allow the stabilization of such small skyrmions. Small skyrmions are generally more stable at higher DMI and lower effective anisotropy [13]. Frustrated exchange, which arises from interface effects, was shown to increase the activation energy for skyrmion collapse in Pd/Fe/Ir(111) [14]. In Rh/Co/Ir(111), skyrmions of 5 nm stabilized at zero magnetic field were recently observed at 4 K in the virgin FM state via spin-polarized STM [1]. A combination of DFT calculations and atomistic GNEB simulations showed that Rh/Co/Ir(111) exhibits a strong exchange frustration, which in turn plays an important role in the stabilization of small skyrmions. Additionally, one may attempt to tune the different system parameters in order to simultaneously increase the energy barrier, and decrease the entropy of the saddle point configuration, thus resulting in a more negative ΔS enhancing the effect of entropic narrowing. This strategy was recently explored in Ref. [15], in which the influence on both the prefactor and the attempt frequency of the extended atomistic exchange parameters, the layer stacking order, and the number of layers was extensively studied. We note that the prefactor of the skyrmion collapse rate has, so far, rarely been calculated in multilayer systems. Skyrmion stability should, in principle, increase in multilayers, for the following reasons: (i) the energy barriers increase if the spin texture is homogeneous along z (for a stacking of the layers along the z-direction) [16]; (ii) if the change in entropy scales like the skyrmion volume in three dimensions [10], then the attempt frequencies may decrease even further. The increase of the lifetime with increasing number of layers was theoretically confirmed in [15]. To this day, the smallest skyrmions observed at room temperature were 30 to 40 nm in diameter, and were reported in multilayer asymmetric stackings of magnetic and non-magnetic heavy metal layers [16, 17].

Nonetheless, predicting skyrmion stability in multilayers is challenging. With the increase in the system size, the computational cost increases as well. Additionally, dipolar interactions should, in that case, be treated more accurately than via an effective anisotropy approximation [18, 19]. Hybrid chirality effects may arise from the interplay of dipolar interactions and DMI [20] and have complex consequences on the thermal stability. For methods based on reaction rate theory, three-dimensional systems with dipolar interactions require the diagonalization of very large, non-sparse matrices. In direct Langevin or path sampling simulations, the cost—and therefore the CPU time—of each iteration increases a lot. These issues will require innovative numerical or theoretical solutions, such as analytical models that may enable the fitting of the ratio of eigenvalues [21] in large systems, combining existing methods, or a different class of methods altogether. In the past, estimating the stability of magnetic structures has often been synonymous with estimating internal energy barriers, while assuming a typically constant value of the prefactor within the gigahertz range [22–25]. Hence, it has been well accepted within the magnetism community that aiming for internal energy barriers of the order of $\approx 50k_BT$ at room temperature is a sufficient and necessary condition in order to achieve the desired 10-year stability of magnetic storage bits. This is no longer valid in the case of magnetic skyrmions—and may, in fact, not be valid in many other magnetic systems relevant for applications [26]—and the search for small, room-temperature stable skyrmions must also necessarily involve some engineering of the energy landscape to tune the Arrhenius prefactor, in addition to the activation barrier.

It is also worth mentioning that antiferromagnetic (AFM) skyrmions have been recently observed in a synthetic antiferromagnet [27] and are currently very promising for applications due to their attractive properties—i.e., fast internal modes in the THz range and vanishing dipolar fields. The latter have been shown to prevent the stabilization of ultrasmall FM skyrmions [28], especially at zero magnetic field [16]. AFM skyrmions are also immune to the skyrmion Hall effect, which deflects the motion of their FM counterparts with respect to any driving force [29], and can lead to their destruction at the sample edges, for instance in racetrack memrories [30]. The stability of skyrmions in an AFM monolayer was investigated theoretically with HTST, but they were so far only found to survive at best for a few milliseconds below 50 K [31].

As for skyrmion-based technological applications, some problems are yet to be tackled. In general, complete theoretical understanding is still lacking concerning skyrmion nucleation by external fields, spin-polarized currents, or laser pulses at room-temperature, and the effect of a driving force on their thermal stability out of thermal equilibrium has not been studied.

In fine, magnetic skyrmions may, or may not, constitute the future of magnetic data storage. Nevertheless, with their unique properties, they remain a rich and fascinating topic for fundamental research.

References

1. Meyer S, Perini M, von Malottki S, Kubetzka A, Wiesendanger R, von Bergmann K, Heinze S (2019) Isolated zero field sub-10 nm skyrmions in ultrathin Co films. Nat Commun 10:3823
2. Hagemeister J, Romming N, Von Bergmann K, Vedmedenko E, Wiesendanger R (2015) Stability of single skyrmionic bits. Nat Commun 6:8455
3. Wild J, Meier TN, Pöllath S, Kronseder M, Bauer A, Chacon A, Halder M, Schowalter M, Rosenauer A, Zweck J et al (2017) Entropy-limited topological protection of skyrmions. Sci Adv 3(9):e1701704

4. Yelon A, Movaghar B (1990) Microscopic explanation of the compensation (Meyer-Neldel) rule. Phys Rev Lett 65:618–620
5. Yelon A, Movaghar B, Branz HM (1992) Origin and consequences of the compensation (Meyer-Neldel) law. Phys Rev B 46:12244–12250
6. Bessarab PF, Müller GP, Lobanov IS, Rybakov FN, Kiselev NS, Jónsson H, Uzdin VM, Blügel S, Bergqvist L, Delin A (2018) Lifetime of racetrack skyrmions. Sci Rep 8(1):1–10
7. Bessarab PF, Uzdin VM, Jónsson H (2012) Harmonic transition-state theory of thermal spin transitions. Phys Rev B 85:184409
8. Desplat L, Vogler C, Kim J-V, Stamps RL, Suess D (2020) Path sampling for lifetimes of metastable magnetic skyrmions and direct comparison with Kramers' method. Phys Rev B 101(6):060403(R)
9. Müller GP, Bessarab PF, Vlasov SM, Lux F, Kiselev NS, Blügel S, Uzdin VM, Jónsson H (2018) Duplication, collapse, and escape of magnetic skyrmions revealed using a systematic saddle point search method. Phys Rev Lett 121(19):197202
10. von Malottki S, Bessarab PF, Haldar S, Delin A, Heinze S (2019) Skyrmion lifetime in ultrathin films. Phys Rev B 99(6):060409
11. Fert A, Reyren N, Cros V (2017) Magnetic skyrmions: advances in physics and potential applications. Nat Rev Mater 2(7):17031
12. Fert A, Cros V, Sampaio J (2013) Skyrmions on the track. Nat Nanotechnol 8(3):152
13. Legrand W (2019) Crafting magnetic skyrmions at room temperature: size, stability and dynamics in multilayers. PhD thesis, Université Paris-Saclay
14. von Malottki S, Dupé B, Bessarab P, Delin A, Heinze S (2017) Enhanced skyrmion stability due to exchange frustration. Sci Rep 7(1):12299
15. Hoffmann M, Müller GP, Blügel S (2020) Atomistic perspective of long lifetimes of small skyrmions at room temperature. Phys Rev Lett 124:247201
16. Moreau-Luchaire C, Moutafis C, Reyren N, Sampaio J, Vaz C, Van Horne N, Bouzehouane K, Garcia K, Deranlot C, Warnicke P et al (2016) Additive interfacial chiral interaction in multilayers for stabilization of small individual skyrmions at room temperature. Nat Nanotechnol 11(5):444–448
17. Soumyanarayanan A, Raju M, Oyarce AG, Tan AK, Im M-Y, Petrović AP, Ho P, Khoo K, Tran M, Gan C et al (2017) Tunable room-temperature magnetic skyrmions in Ir/Fe/Co/Pt multilayers. Nat Mater 16(9):898
18. Winter JM (1961) Bloch wall excitation. Application to nuclear resonance in a bloch wall. Phys Rev 124:452–459
19. Lobanov IS, Jónsson H, Uzdin VM (2016) Mechanism and activation energy of magnetic skyrmion annihilation obtained from minimum energy path calculations. Phys Rev B 94(17):174418
20. Legrand W, Chauleau J-Y, Maccariello D, Reyren N, Collin S, Bouzehouane K, Jaouen N, Cros V, Fert A (2018) Hybrid chiral domain walls and skyrmions in magnetic multilayers. Sci Adv 4(7):eaat0415
21. Fiedler G, Fidler J, Lee J, Schrefl T, Stamps RL, Braun H, Suess D (2012) Direct calculation of the attempt frequency of magnetic structures using the finite element method. J Appl Phys 111(9):093917
22. Weller D, Moser A (1999) Thermal effect limits in ultrahigh-density magnetic recording. IEEE Trans Magn 35(6):4423–4439
23. Chen E, Apalkov D, Diao Z, Driskill-Smith A, Druist D, Lottis D, Nikitin V, Tang X, Watts S, Wang S et al (2010) Advances and future prospects of spin-transfer torque random access memory. IEEE Trans Magn 46(6):1873–1878
24. Lederman M, Schultz S, Ozaki M (1994) Measurement of the dynamics of the magnetization reversal in individual single-domain ferromagnetic particles. Phys Rev Lett 73(14):1986
25. Cortés-Ortuño D, Wang W, Beg M, Pepper RA, Bisotti M-A, Carey R, Vousden M, Kluyver T, Hovorka O, Fangohr H (2017) Thermal stability and topological protection of skyrmions in nanotracks. Sci Rep 7(1):4060

26. Desplat L, Kim J-V (2020) Entropy-reduced retention times in magnetic memory elements: a case of the Meyer-Neldel compensation rule. Phys Rev Lett 125:107201
27. Legrand W, Maccariello D, Ajejas F, Collin S, Vecchiola A, Bouzehouane K, Reyren N, Cros V, Fert A (2020) Room-temperature stabilization of antiferromagnetic skyrmions in synthetic antiferromagnets. Nat Mater 19(1):34–42
28. Büttner F, Lemesh I, Beach GS (2018) Theory of isolated magnetic skyrmions: from fundamentals to room temperature applications. Sci Rep 8(1):1–12
29. Nagaosa N, Tokura Y (2013) Topological properties and dynamics of magnetic skyrmions. Nat Nanotechnol 8(12):899–911
30. Zhang X, Zhou Y, Ezawa M (2016) Antiferromagnetic skyrmion: stability, creation and manipulation. Sci Rep 6:24795
31. Bessarab PF, Yudin D, Gulevich DR, Wadley P, Titov M, Tretiakov OA (2019) Stability and lifetime of antiferromagnetic skyrmions. Phys Rev B 99:140411

Appendix A
Numerical Accuracy in Langer's Theory Implementation

A.1 Checking for Numerical Noice: Influence of the Lattice Size

It should be noted that numerical rounding errors can affect the accuracy of the ratio of eigenvalues, as previously mentioned in Ref. [1]. In Table A.1, we gather results of simulations performed for different lattice sizes. For processes involving a single skyrmion and 30×30 simulated sites, the skyrmion is confined by the boundary and less stable. However, at 50×50 sites and above, we observe very little variations in the different terms of the prefactor. In the case of the two-skyrmion process, the rate loses its lattice-size dependence for 70×70 simulated sites and above. If significant numerical noise were contained within the ratio of eigenvalues, the prefactor would change when varying the number of simulated sites in the system. This does not seem to be the case here, which increases confidence in the present results.

A.2 Influence of the Tolerance on the GNEB Force

When the order of the saddle point involved in a particular mechanism is higher than one, the CI-GNEB scheme will usually make the path change to a different path involving a first-order SP. This is the case even when the path was previously well relaxed to a low tolerance by the GNEB scheme without a climbing image. This occurs because the CI is only designed to climb uphill along the reaction coordinate. If other degrees of freedom are unstable at the SP, the CI will move downhill along these directions, and will therefore usually not converge to a higher-order SP. However, we might still want to estimate the contribution of paths involving higher-order SPs to the total rate. In that case, we have to set a lower tolerance on the GNEB force, so that the scheme will terminate before the path changes. In Table A.2 we compare the energy barrier and the terms of the prefactor for low and high tolerance on the

L. Desplat, *Thermal Stability of Metastable Magnetic Skyrmions*, Springer Theses,
https://doi.org/10.1007/978-3-030-66026-0

Table A.1 Terms of the rate prefactor calculated for different number of sites $N = N_x \times N_y$ (here $N_x = N_y$)

(a) Collapse (mechanism n°1)				
N	ΔE (J_{ex})	$\Omega_0(\times 10^{-5})$	λ_+ (GHz)	f_0 (MHz)
30 × 30	2.76	3.79	1198.73	7.24
50 × 50	2.83	3.51	1200.47	6.70
70 × 70	2.83	3.49	1200.48	6.67
(b) Escape (flat boundary)				
N	ΔE (J_{ex})	$\Omega_0(\times 10^{-2})$	λ_+ (GHz)	f_0 (GHz)
30 × 30	3.23	1.20	521.60	3.80
50 × 50	3.28	1.24	522.94	4.14
70 × 70	3.28	1.24	522.94	4.13
(c) 2 sk. collapse				
N	ΔE (J_{ex})	$\Omega_0(\times 10^{-5})$	λ_+ (GHz)	f_0 (MHz)
50 × 50	2.82	1.86	1198.11	3.56
60 × 60	2.82	2.49	1200.22	4.76
70 × 70	2.82	2.33	1200.23	4.44
80 × 80	2.82	2.32	1200.23	4.43

Table A.2 For collapse mechanism n°1, energy barrier, terms in the rate prefactor and lifetime comparison for different values of the set tolerance on the (CI)-GNEB force

F_{tol}	ΔE (J_{ex})	$\Omega_0(\times 10^{-5})$	λ_+ (GHz)	f_0 (MHz)	τ (300 K) (ms)
1×10^{-12}	2.8256343320	3.51	1200.473	6.70	8.2
6×10^{-5}	2.8256343324	3.76	1200.474	7.18	7.7

GNEB force in the case of collapse mechanism n°1. Although we lose some accuracy in the rate, the results still remain close so that, on the scale of skyrmion stability, the skyrmion lifetime obtained at higher tolerance is still acceptable. Alternatively, one could interpolate the energy in the vicinity of the SP with a high enough density of interpolation points, and choose the highest energy point as the SP.

Reference

1. Fiedler G, Fidler J, Lee J, Schrefl T, Stamps RL, Braun H, Suess D (2012) Direct calculation of the attempt frequency of magnetic structures using the finite element method. J Appl Phys 111(9):093917